SDGs and Textiles

Series Editor

Hafeezullah Memon ⓘ, International Institute of Silk, Zhejiang Sci-Tech University, Hangzhou, China

The book series "SDGs and Textiles" addresses the strategies to achieve sustainable development goals (SDGs) in the present, past, and future. It presents books about the present and future policies of textile ministries of different countries, and books related to sustainability education around different parts of the world in the textile sector. Moreover, it would welcome the conference proceeding related to SDGs and Textiles. The series would cover books comparing the sustainability and SDGs of different institutions and countries. The individual book volumes in the series are thematic. The goal of each book is to give readers a comprehensive overview of a different area of sustainability in the textile sector. As a collection, the series provides valuable resources to a broad audience in academia, the research community, industry, and anyone looking to expand their knowledge of SDGs and Textiles.

Textiles and life are together – life cannot be separated from textiles as it is the most important need for human beings after food. In 2015, the United Nations General Assembly proposed 17 interlinked global goals to be achieved by 2030. Since then, academia and industry have paid much attention to achieving these goals. Textile found its close relation with almost all of these 17 goals.

SDG 1 - No Poverty: Poverty would never be overcome by a charity only; it is essential to develop people's skills to have a better and wealthy life. Thus, the textile can be considered an excellent discipline to achieve this goal by creating jobs and small and medium businesses.

SDG 2 - Zero Hunger: Through the effective utilization of advanced application of Agrotech Textiles, it is possible to have higher crop yields and save crops from rough weather, unexpected rains, floods, insects, etc.; thus, geotextiles play an essential in achieving this goal of sustainable development.

SDG 3 - Good Health & well-being: There has been much health consciousness after Covid19, and medical textiles assist in getting good health and well-being.

SDG 4 - Learning & Education: Textile or fashion has remained a significant discipline for societies for ages, and there has always remained much to explore in this field. Textile-related universities may play a vital role by offering free access to their education resources, training and spreading information among the locals.

SDG 5 - Gender Equality: The textile sector is one of the industrial sectors that accepted gender equality long ago; in particular, the garment sector has more females than males. Thus, the textile sector has been doing gender equality. Moreover, there has been a recent trend for Gender Neutral Clothing, which need worth studying and may further assist gender equality.

SDG 6 - Clean Water & Sanitation: Textiles could be achieved through filtration, and of course, textile is one of the critical materials for filtration.

SDG 7 - Affordable & Clean Energy: With the recent advancement in material science and engineering, the textile sector has come on the front for, not only by using this clean energy during textile production but also by assisting the production of this clean energy, either in the form of wind turbines blades made of textile composites or by energy harvesting from T-Shirts, etc.

SDG 8 - Decent Work: Recently, there has been much attention that the textile workers are not paid well, labor rights are not cared about, etc.

SDG 9 - Industry and innovation: Textile Industry always follows innovation; the textile companies that do not chase innovation cannot survive in the market.

SDG 10 - Reduced Inequalities: Getting better life and well-being would help reduce inequalities in the textile industry.

SDG 11 - Sustainable Cities: Sustainable Textile Cities through Buildtech and transport textiles.

SDG 12 - Consumption and Production: Textile and garment consumption and production all come under.

SDG 13 - Climate Action: Oekotech or Ecotech Textile, waste management of textiles are upfront to achieve this goal of sustainable development.

SDG 14 - Life Below Water: Mitigating microfiber waste in rivers and oceans may come under the context of it. There has been much attention on this subject after passing the bill at the parliament level of the UK.

SDG 15 - Life on Land: Geotech or Geotextiles studies life on land.

SDG 16 - Peace, Justice, and Strong Institutions: Protective textiles are doing their best to achieve peace, justice, and strong institutions.

SDG 17 - Partnerships for the Goals: The application of textiles to achieve sustainable development goals is only an example. In all textiles sectors, combined efforts of all the goals are essential to achieve true sustainability.

Manuel Gausa · Giorgia Tucci

Knitting Food: Food and Eco-textiles

New Perspectives for Sustainable Urban Production Systems

 Springer

Manuel Gausa
Department of Architecture and Design
University of Genoa
Genoa, Italy

Giorgia Tucci
Department of Architecture and Design
University of Genoa
Genoa, Italy

ISSN 2948-1236   ISSN 2948-1244 (electronic)
SDGs and Textiles
ISBN 978-981-97-7581-1   ISBN 978-981-97-7582-8 (eBook)
https://doi.org/10.1007/978-981-97-7582-8

This Springer imprint is published by the registered company Springer Nature Singapore Pte Ltd.
The registered company address is: 152 Beach Road, #21-01/04 Gateway East, Singapore 189721,
Singapore

If disposing of this product, please recycle the paper.

Preface

According to the FAO Reports on *The State of Food Security and Nutrition in the World* (2021 edition) [1], various factors have contribute negatively to difficult the goal of ending global hunger and malnutrition by the 2030s: the complications derived from the COVID-19 pandemic have been added to the economic effects derived from the current war escalations and an inflationary imbalance that have forced to made new estimates on forecasts of food costs and the capability to facilitate an universal access to decent, safe and healthy diets.

Hunger and malnutrition have continued to reach such critical levels that they force to generate new reflections on the own world situation, food security and the general right to proper nutrition. The determining parameters of a deficit trend, the frequency and intensity of which tends to increase, should not only be associated with the increase in diseases and conflicts but, also, with a growing variability in the extreme conditions produced by climate change.

The economic slowdowns have also aggravated the poverty thresholds and the levels of inequality, exaggerated and persistent. Millions of people in the world continue to suffer from food deficits and different forms of malnutrition because they cannot afford the cost of minimally healthy diets [2].

Paradoxically, and at the same time—especially in the most developed countries— huge amounts of food continue to be wasted in the world. In the European Union (EU) alone, an average of 87.6 million tons of food has been wasted in recent decades, which is equivalent to approximately 173 kilograms per person. Most of the waste, according to data from the European Commission itself, is concentrated in homes and domestic habits, followed immediately by the industrial processes of food processing and catering, at all levels and scales [3].

Loss and waste of food in any case exacerbate the risk of food insecurity and malnutrition, excessive use of water and irrational consumption of energy sources, at a time when environmental risks are increasing in parallel with hunger in the world. The EU countries, also committed to the United Nations Sustainable Development Goal (halving, by 2030, per capita food waste), have recently adopted concrete measures to prevent and reduce such losses, as well as to reuse or recycle food waste for other purposes.

The EU Framework Directive on Food Waste requires Member States to:

- reduce the amount of food lost during production and distribution;
- reduce food waste at home;
- encourage rational consumption and donation of food;
- monitor and evaluate the implementation of EU measures on the prevention of food waste;
- promote the reuse of food surpluses as feed or compost materials or their recycling (or second life) to generate new materials and products [3].

With the implementation of the *European Green Pact* in December 2019, the EU reaffirmed this commitment, establishing a series of policies and instruments as part of the Action Plan for the Circular Economy and the Strategic Line "From Farm to Fork" (associated with the *Cradle to Cradle* and *Km0 Circuits* concepts), two fundamental pillars of the Green Pact.

The promotion of various platforms of the EU itself and of parallel projects (often financed with competitive funds) seeks to share the best practices generated in this direction and evaluate the progress made over time.

In this sense, this publication is presented as a choral essay on the different links that can be generated between innovative territorial planning, agriculture, landscape and functional mix-use; but also, between food production and consumption and new research dynamics around the real capacity to generate efficient bio-materials derived from the agro-surplus and/or food-waste themselves.

Conceived, promoted and signed by the GIC-Lab-UNIGE research group (Università degli Studi di Genova) the contents are presented through the following format:

a) a first overview (1. Rural Landscapes and Agro-Revolutions) on the evolution of food production and the mobility-transportation of agro-productive systems; on the need to identify new prospective and integrated land-landscape-agriculture approaches; as well as on the importance of new mixed (eco-sensitive and techno-efficient) approaches in the field of second-life food and its optimisation through innovative recycling methodologies;

b) a second part (2. Performative Food-Matters: Food, Waste and Biomaterials), conceived as a reasoned analysis of bioproducts derived from food by-products and the circular reconversion of waste into resources;

c) a third part (3. Food Second-life as Innovative Research) completes the analysis framework with international references of good practices, experiments and experiences developed in recent years in biomaterials research derived from food waste. Specifically, the results of the Creative Food Cycle (CFC) project, developed within the Creative Europe programme, are outlined;

d) a fourth part (4. Ecotextiles as (Cr)edible Matters) frames eco-textiles as key topic of this study, focusing on the new challenges of the textile industry, from sustainability processes to advanced technology, from new eco-fabrics to circular design.

e) a fifth part (5. Knitting-Food: a Global Catalogue of Innovative Eco-textiles) is presented as an Atlas-catalogue of good practices: projects, experiments, prototypes or internationally trade-marked products and case studies of key experiences.

f) a sixth and concluding part (6. Eco-Textiles: New Challenges and Future Perspectives) summarises the arguments presented, integrating considerations on regenerative agriculture and "transitioning design" that complement the previous reflections and look to future perspectives.

This study can be interpreted as an initial global reflection and overview on the equation "*Food + Waste + Agro-production + Tecno-innovation*" and its relationship with eco-textiles as new *Food-Matters*.

The possibility of offering a new, more proactive and innovative framework for action in the field of eco-textiles, in which the old conservative connotations of the term sustainability can be combined with more active and creative ones; more positive (and pro-positive) and innovative processes, linked to the terms combination, adaptation, conjugation, hybridisation [4].

This publication is not intended to be an exhaustive review of the topic but to encourage an initial critical and functional reading, based on the exchange of experiences and training, education and R&D activities promoted by the GIC-Lab.

We also thank all the participants who have made these reflections possible (designers, start-ups, researchers, tutors, students, members of the CFC consortium, cultural associations and research centres, experts and companies involved) for their generous contribution and proactive energy.

Genoa, Italy   Manuel Gausa
                                                                                                      Giorgia Tucci

References

1. FAO, IFAD, UNICEF, WFP and WHO. 2021. The State of Food Security and Nutrition in the World 2021. Transforming food systems for food security, improved nutrition and affordable healthy diets for all. Rome, FAO. https://doi.org/10.4060/cb4474en
2. FAO, IFAD, UNICEF, WFP and WHO. 2023. The State of Food Security and Nutrition in the World 2023. Urbanization, agrifood systems transformation and healthy diets across the rural–urban continuum. Rome, FAO. https://doi.org/10.4060/cc3017en
3. European Consilium, Reducing food loss and food waste. https://www.consilium.europa.eu/en/policies/food-losses-waste/
4. Pericu S., Gausa M., Tucci, G., Ronco Milanaccio, A. eds. (2021), Creative Food Cycles Experience. Goa CFC-festinar: a virtual banquet for an innovating research celebration. Genova: ADDDoc Logos. https://gup.unige.it/sites/gup.unige.it/files/pagine/Creative_food_cycles_experience_ebook_indicizzato.pdf

Acknowledgements

Knitted Food: food and eco-textiles. New perspectives for sustainable urban production systems is the result of research shared by both authors, however, in Chapters 1. "Rural Landscapes and Agro-Revolutions", 4. "Eco-Textiles as (Cr)edible Matters" and 6. "Eco-Textiles: New Challenges and Future Perspectives" there is a greater contribution from Manuel Gausa, while in Chapters 2. "Perfomative Food-Matters: Food, Waste and Biomaterials", 3. "Food Second-Life as Innovative Research" and 5. "Knitting-Food: a Global Catalogue of Eco-Textiles" have a major contribution from Giorgia Tucci.

We also would like to thank:

- The members of GicLab-UNIGE—Nicola Canessa, Emanuele Sommariva, Centanaro—for their involvement in the research activities during these years;
- Silva Pericu, Joërg Schroëder, Areti Markopoulou for their contribution to the Creative Food Cycles research;
- Chiara Maresca and Giulia Geri for their participation in the catalogue work;
- All the designers and creatives mentioned in Chap. 5, who made their material available and kindly consented to its dissemination.

The project descriptions in Chap. 5. "Knitting-Food: a Global Catalogue of Eco-Textiles" are elaborations by the authors based on information gathered from the designers' websites and the information material provided.

Contents

Chapter 1
Rural Landscapes and Agro-revolutions

1.1 Agriculture, Food and Primary Revolution (Foundational)

In the well-known and brilliant illustrations that illuminate the *Livre des Heures du Duc de Berry* (Herman, Paul and Johan Limbourg, 1410), one can appreciate—always placed in the background—the walled city (a solid and compact object from which emerge its "symbolic" constructions, temples, palaces, towers, etc.) and, in the foreground, the "countryside": the cultivated territory outside the walls, the agrarian space destined to feed not only the farmers who work it but the entire urban and rural population.

Urb versus *Rur*, two complementary yet clearly opposed worlds.

For a long time, the products of agriculture were conceived solely as food—nourishments and/or nutrients—necessary for vital energy, human and animal.

Occasionally, they are also used as fertiliser or as fuel from waste.

The "domestication" of plants and animals, as well as the development and transmission of techniques to cultivate or breed them in a productive way, is at the very basis of the history and evolution of agriculture and its main mission: food.

From about 20,000 B.C., human beings collected (and thus began to gather) wild grains for their sustenance, and from about 10,000 B.C., the eight "founder crops" began to develop in the Mediterranean Levant, i.e. the first Neolithic species systematically sown: three cereals (spelt, wheat and barley), four legumes (peas, lentils, chickpeas and broad beans) and one fibre (flax).

Almost in parallel, another major nutritional component, rice, was domesticated in China between the eleventh and sixth centuries B.C., followed by soybeans and other various types of beans [1].

In the Middle Ages—both in the Islamic world and in Europe—agriculture had already been substantially transformed with the use of successively improved techniques and the expansion of more varied crops such as barley, sugar cane, cotton, onions, garlic and turnips, fruit trees (with oranges in the first place), wine and grapes or various varieties of rice. The importance of horses, donkeys and cattle was also commonly combined with the grazing of sheep and goats.

After the discovery of America in 1492, the exchange between continents brought important crops from America to Europe (such as tomatoes, peppers, beans, potatoes, sweet potatoes, pumpkins, corn and sunflowers) and, in turn, crops and cattle were exported from the Old World to the New World.

With the onset of the industrial revolution, agriculture grew at a much faster pace—especially in developed countries, but also, to a lesser extent, in developing countries—experiencing strong growth in production due to the consolidation of the use of specialised machinery and the resulting mechanisation [1] (Fig. 1.1).

1.2   Agriculture, Food and Secondary (Industrial) Revolution

The development of the industrial revolution was to bring with it, then, a renewed boom in agriculture and the accelerated transition from the more artisanal (or manual) modes of production to a new type of production (industrial) based on new mechanical processing systems, new sources of energy, new types of cultivation (with the incorporation of synthetic fertilisers, pesticides, selective breeding, etc.) and a clear division and seriation of labour, among other factors.

Gasoil-powered tractors (developed in the first two decades of the twentieth century) were soon to become a generalised work tool, favouring the replacement of traditional draft animals (particularly horses and oxen) with increasingly specialised machinery and devices (mechanical harvesters, seeders, pruners, transplanters, etc.) that were to considerably increase the speed, volume and scale of agricultural production [2].

From the mid-twentieth century onwards, modern scientific research would also accelerate genetic and transgenetic engineering and the development of new biofuels such as ethanol as well as the use of nitrogen fertilisers.

From the first foods processed by rudimentary and artisanal (local) methods, humanity would end up consolidating and/or prioritising, more and more, those methods of mass production (global) capable of ensuring a greater quantity of fresh foods and also (and increasingly) those preserved and/or precooked.

Modern agriculture associated with the industrial era would therefore produce more "efficient" species and larger cultivated areas, but would also give rise to many of the frequent environmental (and also social, political and economic) problems resulting from the combination of the primary and secondary sectors: pollution, CO_2 emissions, water contamination, energy problems associated with fuels pollutions or questions about increasingly genetically modified *organisms*; but also unemployment, intensive crops, the need for agricultural subsidies, etc.

Fig. 1.1 Different illustrations of the *Livre des Heures du Duc de Berry* (Herman, Paul and Johan Limbourg, 1410)

The soon-to-be-named food industry, which emerged from the industrial revolution itself, would seek to address in an efficient, vocationally rational way— but also developmental, from a material and productive point of view—all the processes related to the so-called "food chain": that is, the different phases of *cultivation, harvest, transport, distribution* and *storage, processing, conservation* and *food services* for human and animal consumption [3].

Fig. 1.2 Waterloo Boy Tractor, John Deere Company, 1917–24. *Source* Agricultural online magazine

Its progressive development would end up affecting, in an evident and notable way, daily nutrition and food itself, by increasing the quantity, variety and diversity of products available in and for the different diets generated.

In fact, the increase in production would in turn lead to a significant expansion in the systems of conservation, preservation and packaging or canning (processed and/or pre-cooked foods), which are increasingly present in social consumption habits, and which contain highly synthetic components in their additive elements and packaging (plastics, cellulose, etc.).

In parallel to these processes, the successive and specific "food laws"—general or particular, depending on the region—try to alleviate the harmful effects through greater regulation and control of the various processes linked to the "food products" themselves, from their production to their transfer, storage or processing (Fig. 1.2).

1.3   Agriculture, Food and Third (Green) Revolution

The universal awareness of "hunger" as a social, sensitive and solidarity issue was expressed in a remarkable way by the growing importance of the social movements that emerged in the second half of the twentieth century, thanks to the development of the mass media and non-governmental organisations. Its effects were to be decisively addressed after the Second World War and in parallel with the creation of international cooperation organisations such as the UN, FAO, etc.) [14].

Between the 1940s and 1970s—and through the so-called *Third Agricultural Revolution* or *"Green" Revolution*—a series of R&D, development and research initiatives were to be prioritised that would exponentially increase global agricultural production to urgently answer to these questions: initiatives involving the development of high-yielding cereal varieties, the expansion of irrigation infrastructure, the modernisation of planting techniques, as well as the adoption of hybrid seeds, synthetic fertilisers and pesticides.

Although the Green Revolution was to achieve significant quantitative success, it was not always to be qualitative: the expansion of low-quality protein and high-carbohydrate cereal varieties, which were widespread and increasingly predominant in the world, was to lead to widespread nutritional impoverishment and many problems associated with the crucial area of health [4].

Indeed, the progressive increase in the world's population over the last century would tend to exacerbate the obvious need to promote the universal right to a decent and varied diet. In many countries, the direct consequence was going to be this stimulus to increasingly massive production, both in the meat and agricultural industries, with a degradation of the living conditions of animals and the interruption of the natural cycle of plants.

The majority of mass-produced fruits, vegetables and cereals would grow through intensified monocultures, a short-term solution that would tend to cause environmental stress situations through strong alterations of the agricultural soil inflicted by the massive depletion of the nutrients necessary for the survival of the plantations.

The crops themselves are often sprayed with harmful pesticides, causing terrible cascading effects through water pollution, indirectly destroying nearby eco-systems [5].

However, a new ecological and sustainable awareness, growing since the 1990s (first Rio Summit in 1992, Kyoto Summit in 2005, second Rio Summit in 2012), was to entail a clear critical reconsideration of the *Green Revolution* itself.

The emergence of a new sensitivity linked to the preservation of the landscape, the rational use of soil and cultivated areas, as well as basic resources such as water or energy systems, made it necessary to seek other types of productivity—accessible, sufficient and balanced—while safeguarding the viability of agricultural and wild eco-systems.

In this sense, the parallel evolution of new technologies would favour this type of approach [6, 7].

The new environmental but also (in parallel) technological revolution—digital, informational, interconnective but above all interactive (declined at all levels, and therefore with the environment itself)—would prove to be a key factor in optimising, through accurate data and efficient formulation parameters, increasingly responsible and responsive responses.

New paradigms such as the intelligent processing and managing processes (*smart*), the circularity of them (*cradle to cradle*), the eco-agriculture, the permaculture, the biodiversity (and the nutritional variety itself) or the "slow food" and the use of "Km 0" resources would be combined with new potentials linked to the recycling or reusing of food itself (or rather, of the waste associated with its production and consumption).

Although many of these premises would seem to tend to "recover" or "vindicate" neo- or para-artisan modes and ways, the great paradox of the technological revolution would be that of facilitating complex informational combinations and holistic socio-technological and environmental conceptions in an interactive (that is, relational) way, global and local, connected with many identity sensibilities and with many diversified exchanges.

In this sense, in contrast to the 'universally transformative' dogma of industrial modernity, this strategic interrelationship—positive rather than positivist—would allow for the interconnection of past and present, tradition and innovation, recovery and creation, in a way that is not only *Sustainable* but also *Nexusustainable* or, if you prefer, *(e)Co-compatible* (Fig. 1.3).

1.4   Agriculture, Food and Fourth Revolution (Sustainable or Smart)

In the last decades, the agricultural sector has experienced, in this sense, a considerable progress thanks, precisely, to the new technological advances and a growing digitalization since the early 1990s. These advances are precisely what have favoured not only the emergence of new products associated with all types of crops, but also better process management, optimizing (through precise data, parameters and indicators combined in new formulas and qualitative combinations) an innovative production at all levels.

The objective of the so-called "Smart Agriculture" (or "Agro-Smart") would be, from the outset, to use, in fact, information and communication technologies (ICT) to collect and identify data, process them, reformulate them and achieve more efficient results, using different types of devices and farming systems combined in a new type of "processing logic" (flexible, adaptive and reactive or responsive) capable of responding in a specific and differential way to various particular situations, facilitating, among other things, a greater rationalisation of the use of resources, water, energy or the emergence of new phytosanitary and environmentally sustainable fertilisers [9].

The *Smart City* concept itself—fundamental at the beginning of the second decade of the twenty-first century—basically alluded to a more reactive and responsible and responsive management of a whole set of data (*Big Data*) and indicators (*Balancing Factors*) recorded and combined in a complex way and in integrated systems, based in turn on a number of subsystems (security, water, health, infrastructure, economy, environment, food, agriculture, land use, etc.) managed in a coordinated manner, to ensure more sustainable development and growth [10].

Within the framework of Smart Cities themselves, Smart Agriculture (rural and urban) would tend to favour a shortening of the supply chain between origin and

Fig. 1.3 Intensive agriculture: aerial view of agricultural greenhouses in Albenga, Liguria, Italy, courtesy of Ph. Luciano Rosso, Albenga

destination by trying to ensure healthier food, using less environmentally aggressive farming methods and encouraging the creation of possible new programmes and mixed economies, more advanced and diversified, associated with the agricultural enclaves themselves [10, 11].

The new *agro-techno*-sustainable revolution would be based, in the first instance, on the capacity to generate greater efficiency and rationality in the processes (differential and eco-systemic) in order to generate "more production with fewer resources" not only promoting positive results capable of satisfying the food demand itself, but also generating new "products" beyond food (from eco-programmes to biomaterials) [12].

This change of approach would not be limited to cultivation processes alone, but would extend to the food processing and distribution sectors, as well as to environmental resilience, health, landscape preservation and operational valorisation, urban horticulture and eco-experiential tourism, digital manufacturing, etc. [13].

In this sense, when we speak of new devices, bioproducts and biomaterials resulting from the equation *food + research*, we could summarise them in two groups

- On the one hand, those elements intended to benefit and optimise agricultural production itself (ecological compounds, controlled seeds and seedbeds, adaptable topo-morphological meshes, thermal-sensitive blankets, intelligent structures for greenhouses, reactive protective bags, tracking, recording and control drones, autonomous vehicles, agricultural robots or *agro-bots*, etc.).
- On the other hand, those derived—as has been pointed out—from agricultural production itself—food products and their complements—and declined beyond their primary definition (food, food) to favour other multiple definitions based on reuse, recycling or the creative and qualitative redefinition of food materials and substances themselves (from bio-cosmetics to bio-plastics, bio-fibres and bio-textiles, celluloses or food based dyes, varnishes and binders, new biodegradable or "edible" packaging, digitised or manufactured bio-design, etc.) (Fig. 1.4).

1.5 Agriculture, Food and Fifth Revolution (Circular)

We would refer, then, to a new "bio-technological" phase with a decisive role in agriculture and livestock farming in the twenty-first century, a role that requires innovative bets, transversal technological transfers and entrepreneurial investments beyond the current conventions.

Some authors might consider this fifth "revolution" a direct derivative of the previous one—an evolution rather than a revolution. Whether we are witnessing a fifth renewal of the production logic itself or a subphase of the aforementioned techno-sustainable revolution, [17] more focused on strategic processes geared towards fostering virtuous circularity and the reuse of land resources is a matter for further investigation.

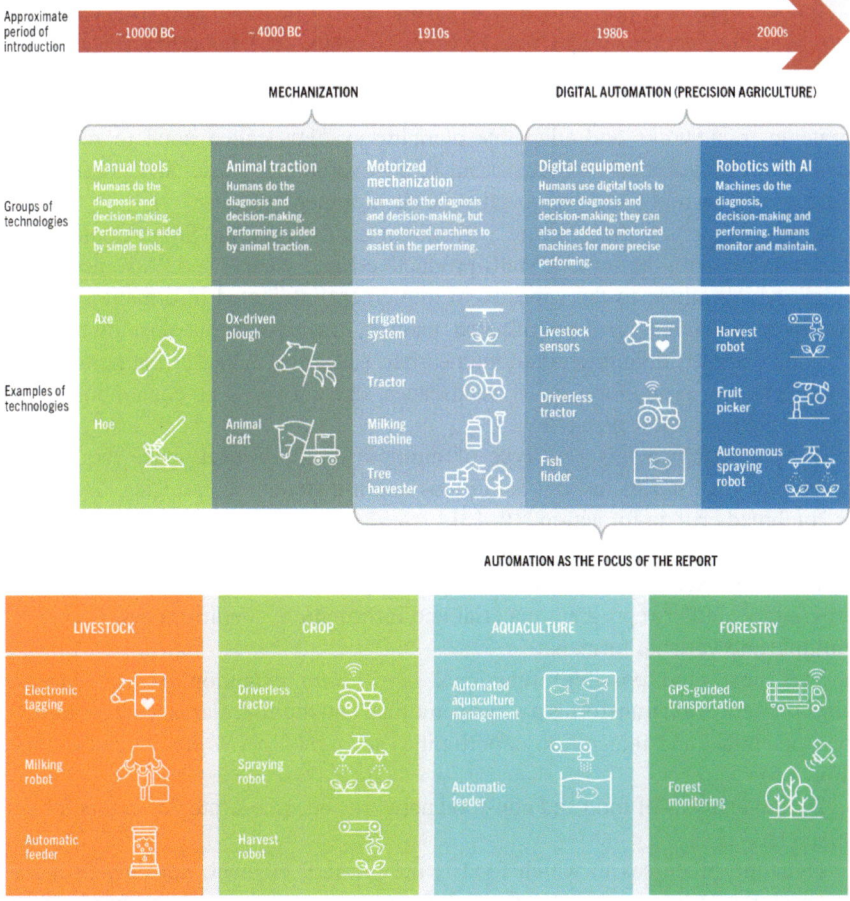

Fig. 1.4 Evolution of agricultural automation and selected digital technologies and robotics with artificial intelligence by agricultural production system. Elaboration FAO, 2022: https://doi.org/10. 4060/cb9479en

In this resilient and intelligent framework—resili(g)ent [18]—urban and interurban agriculture can contribute to healthier and more efficient nutrition processes, through an algorithmic optimisation of environmental and economic parameters, capable of managing: energy and waste cycles, water and material consumption, better management of environmental needs, etc.

New multi-level and combinatorial approaches oriented towards the creation of agro-landscapes not only for food, but also for recreation, through a new structural and reactive projection of pre-existing environmental and socio-cultural conditions [10–13].

Indeed, current approaches tend to favour multi-functional scenarios capable of providing not only resources (food, forest floral crops, etc.), but also services associated with tourism, bio-material production, agro-technological research, recreation, etc. Territorial characteristics, activities and values can be incorporated into new strategic frameworks and adaptable planning systems called upon to sensitively accommodate different, coherent (and often overlapping) uses and programmes, generating new multi-sectoral sustainable development models.

Concepts such as "Smart Cities", "Resilient Landscapes" or "Smart Contexts" are thus coupled with the terms "Multi-productive Landscapes" and "Advanced Planning", which allude to the capacity to implement integrated information systems aimed at managing, in a coordinated manner, the sustainable development and growth of these urban/interurban/multi-urban scenarios linked to the new interactions between agriculture, cities, landscapes and networks as contemporary and paradigmatic relationships [18].

The approach to this new type of multiple or multi-level space requires the development of n-dimensional definitions and n-differential strategies, understood as combined criteria for action [10–13].

The digital world and information technologies have exponentially increased the potential for exchanges between needs and demands, but also their very ability to programme and/or reprogram material and information, conditions and stresses, in multiple and variable formats.

This new (and possible) functional *agro-urban* evolution of the *"rurality-urbanity"* equation transforms a new operational and multi-scalar landscape, through new potentials and integration opportunities for *agro-, peri-, para*-urban contexts (Fig. 1.5).

Digital analysis of food and non-food networks become strategic vectors for new mixed land uses.

Dynamic processes in which food is the primary element, not only as a food material, but as a *hyper-matter*, through the recovery and reuse of by-products and food waste, are capable of being transformed into a new resource thanks to the technological-creative capacities of fab-labs, start-ups, new creative incubators, etc. [8–11].

These new global dynamics and needs have inspired multiple international researches that have developed urban perspective projects in which agriculture, food, land use, information and digitisation are combined in new mixed *hyper-agricultural* scenarios in which food represents a new *hyper-matter* [8–11].

Indeed, unlike the priority given to the "optimization" and "management" of informational data and contextual requests, in order to achieve more rational achievements in crops and productions, in this new stage (which is the one that forms a large part of the interests of this publication), management efficiency would give way to a proactive and reactive/creative action, in which the very notion of "food" as a product would take on a more hybrid and mixed condition in its possible derivations associated with bio-material dynamics, clearly circular.

A close relationship between the terms of the equation 'BIO-technology + AGRO-culture + SOS-tenability + HYPER-matter' would help, on the one hand, to precisely

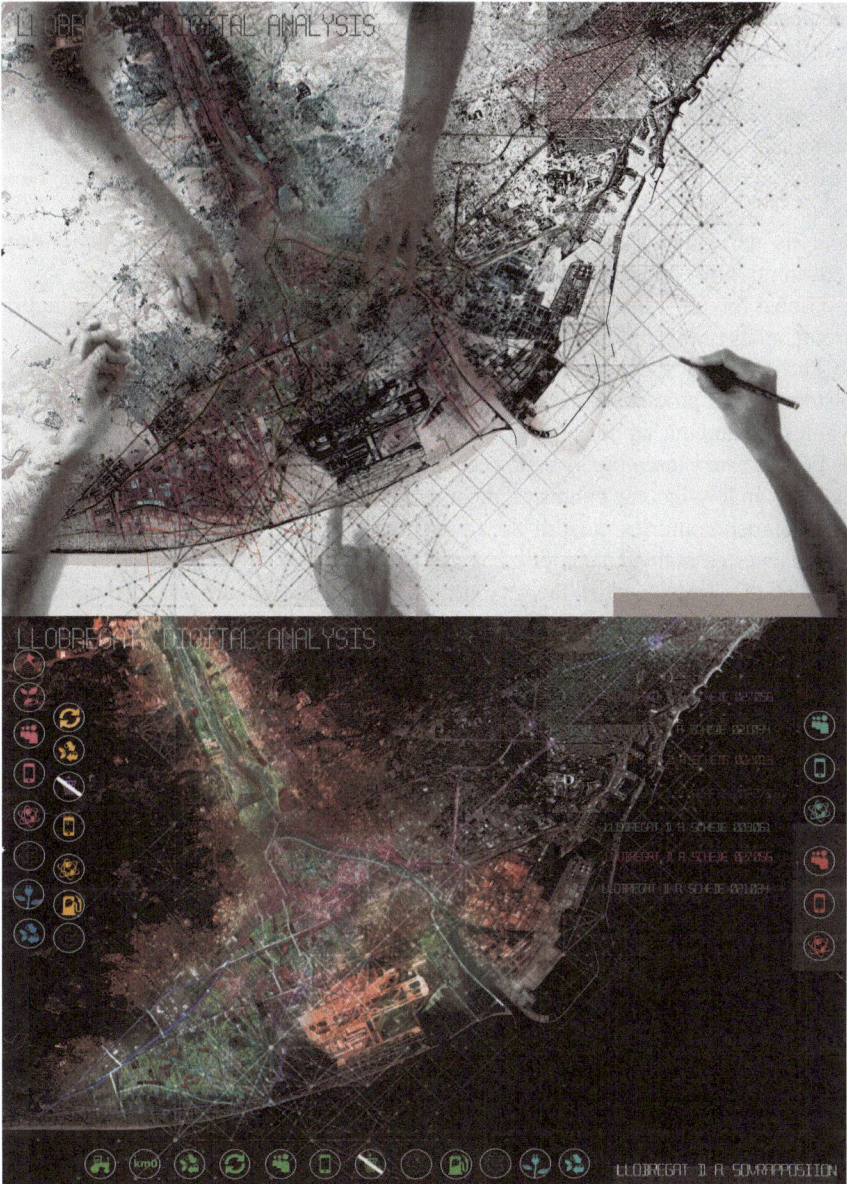

Fig. 1.5 GICLab-Unige (2016–17). LLobregat Agro-Park (Barcelona Metropolitan Area)

and positively increase production per unit of local cultivated area through intelligent development strategies and, on the other hand, to reduce the use and consumption of resources such as soil, water, energy and harmful elements such as fossil fuels, pesticides and fertilisers, favouring product quality and preserving areas of high ecological value.

But, above all, it would call for reconsidering the use of food surpluses and reducing post-harvest losses, reducing, rethinking and/or recycling food waste itself (and its derivatives) and diversifying the use(s) of agricultural production and consumption for other destinations.

Global food loss and waste are one of the key issues today and one of the central protagonists of this publication.

The causes of food waste or loss are numerous and occur throughout the food system, during production, processing, distribution but also during retail (rather than wholesale) and during consumption and its leftovers or waste.

Food waste causes the loss between one-third and one-half of all food produced globally. In low-income countries, most of the losses occur during production, while in developed countries most of the food (about 100 kg (220.5 lbs.) per person per year) is wasted at the consumption stage [15].

To all this must be added the very strong impact of packaging and food packaging waste, which is often associated with the increasing impact of plastics and synthetic or non-degradable elements on the environment: food waste is one of the most important causes of the current environmental impact of agriculture.

The United Nations Food and Agriculture Organizations estimated in 2014 that food waste involved a global economic, environmental and social cost of about $2.6 trillion per year and was responsible for 8 per cent of global emissions of carbon dioxide and 10 greenhouse gases associated with climate change [14–19].

Food waste that is not properly handled or recovered (through composting techniques or biodegradable processes) has decidedly negative environmental effects: landfill gas from anaerobic organic matter is, for example, a major source of methane (a gas with strong impacts on the ozone layer).

As we have noted elsewhere, "food waste reduction" is, today, clearly identified as a decisive part of the necessary progress towards a more sustainable economy, in line with UN Sustainable Development Goal 12, which seeks to *"Halve, within a maximum of one to two decades, global per capita food waste"* (Fig. 1.6).

Fig. 1.6 Smart Agriculture: phytosanitary drones, 2023, CC

References

1. Beadle GW (1980) The ancestry of corn. Sci Am 242(1):112–119
2. Jideani AI, Mutshinyani AP, Maluleke NP, Mafukata ZP, Sithole MV, Lidovho MU, Ramatsetse EK, Matshisevhe MM (2020) Impact of industrial revolutions on food machinery—an overview. J Food Res 9(5):42–52. https://doi.org/10.5539/jfr.v9n5p42
3. Erickson DR (1990) Edible fats and oils processing: basic principles and modern practices. In: World conference proceedings. American Oil Chemists' Society, Champaign
4. Borlaug N (1996) Talk transcript, 34th convocation of the Indian Agricultural Research Institute: New Delhi. https://archive.org/details/BorlaugIARI/page/n3/mode/2up
5. Farinea C (2021) Integrated approach to urban food production. In: Pericu et al (eds) Creative food cycles experience. Goa CFC-festinar: a virtual banquet for an innovating research celebration. ADDDoc Logos, Genova, pp 53–60
6. Mann C (1997) Reseeding the green revolution. Science 277:1038–1043
7. Mann C (1999) Crop scientists seek a new revolution. Science 283:310–314

8. Gausa M (2021) CFC—multiscalar challenges. In: Pericu S, Gausa M, Tucci G, Ronco Milanaccio A (eds) Creative food cycles experience. Goa CFC-festinar: a virtual banquet for an innovating research celebration. ADDDoc Logos, Genova, pp 13–52

9. Buonanno D (2012) Nuovi Paesaggi, Interventi di Rinaturalizzazione Urbana. Planum J Urban 2(25)

10. Tucci G (2020) MedCoast AgroCities. New operational strategies for the development of the Mediterranean agro-urban areas. Trento-Barcelona, ListLab

11. Gausa M, Canessa N, with Tucci G (ed) (2018) Agro-cultures, agro-cities, eco-productive landscapes. Actar Publishers, Barcelona-New York

12. Gausa M, Vivaldi J (2021) The threefold logics of advanced architecture. Actar Publishers, New-York

13. Sommariva E (2015) Cr(eat)ing City. Agricoltura urbana. Strategie per la città resiliente. Trento-Roma-Barcelona, List Lab

14. FAO (2019) The state of food and agriculture 2019. Moving forward on food loss and waste reduction. Rome. https://doi.org/10.4060/CA6030EN

15. United Nations Environment Programme (2021) Food waste index report 2021. Nairobi. https://www.unep.org/resources/report/unep-food-waste-index-report-2021

16. FAO (2011) Global food losses and food waste—extent, causes and prevention. Rome. https://www.fao.org/3/mb060e/mb060e.pdf

17. Kummua M, De Moel H, Porkka M, Siebert S, Varis O, Ward PJ (2012) Lost food, wasted resources: global food supply chain losses and their impacts on freshwater, cropland, and fertiliser use. Sci Total Environ 438:477–489. https://doi.org/10.1016/j.scitotenv.2012.08.092

18. Gausa M (2020) Resili(g)ence, intelligent cities, resilient landscapes. Actar Publishers, New York-Barcelona

19. FAO (2022) The state of food and agriculture 2022. Leveraging automation in agriculture for transforming agrifood systems. Rome, FAO. https://doi.org/10.4060/cb9479en

Chapter 2
Performative Food-Matters: Food, Waste and Biomaterials

2.1 Food Waste Versus Environmental Sustainability

The term "Food Waste" today is a very recurrent word in our vocabulary, social media and language. There is not a real definition in institutional and scientific circles, but an initial clarification on food waste has been given by the FAO.

"Wholesome edible material intended for human consumption, arising at any point in the food supply chain (FSC) that is instead discarded, lost, degraded or consumed by pests" [1].

Food waste refers to food that is fit for human consumption but is discarded, and this occurs throughout the entire supply chain (from producer to consumer): to be precise, from agricultural production to food processing, transport and sale, and finally to use in our homes. Nowadays, all along the global scale, supply chains are getting longer and even more complex:

- Consumers have higher expectations for diversity and convenience of choice.
- More and more people migrate from rural areas to urban centres.
- The distance between the place of production and the place of consumption has increased, making the structure of food distribution and supply increasingly complex.
- Increasing demand for meat, fruit, vegetables and other perishable products increases the risk of loss and waste.

Food waste occurs mainly, but not entirely, at the consumption level, usually related to consumer behaviour determined by policies and regulations, and represents a serious social, economic and ecological "disease" that concerns the Planet we live in, for three main reasons.

1. Economic: every day many people die of starvation or malnutrition, and this is unthinkable, knowing that one third of the world's food is abandoned in the exchanges or is discarded.

M. Gausa and G. Tucci, *Knitting Food: Food and Eco-textiles*, SDGs and Textiles, https://doi.org/10.1007/978-981-97-7582-8_2

2. Environmental: every product produces CO_2 during its life cycle, in addition to which it consumes an excessive amount of water.
3. Psychological: people lose awareness of the value of food.

However, there is a different classification, provided by the FAO [2], which can better define the type of food waste:

- Food Losses: it encompasses food losses that occur upstream in the agro-food chain, i.e. in the production, harvesting, storage and processing phase, of edible parts of plant or animal origin, produced for human consumption. Losses are mainly due to climatic and environmental factors and accidental causes, linked to the limitations of the agricultural techniques used and the infrastructure. In addition to these, there are also losses due to economic reasons, i.e. aesthetic and quality standards imposed by market and food regulations, as well as more or less convenience in harvesting operations [2].
- Food waste also refers to the waste that occurs at the point of distribution at the level of consumers and retailers. This includes conscious choices, i.e. discarding and "throwing away" food that is still fully edible [3].

Food losses depend on logistical and infrastructural limitations, while food waste depends on behavioural factors; this is the main aspect that differentiates them [3].

Household food waste refers to all food that is "lost" in the refrigerator or thrown in the bin before or after consumption. In particular, this occurs in rich countries, whereas in developing countries, waste is zero. Losses are concentrated during the production and storage of food, i.e. in the intermediate stages. Losses and wastage have led to a substantial reduction in the amount of food actually available for human consumption.

Each stage of the agri-food supply chain consists of several operations, both agricultural and industrial, and consequently different types of losses and wastage occur throughout the process. More precisely, six stages of the supply chain have been identified:

1. Sowing, agricultural production and harvesting.
2. First processing.
3. Industrial processing.
4. Distribution.
5. Catering.
6. Domestic consumption.

The first phase of the chain includes those activities closely related to sowing and agricultural production, during which losses may occur because crops are not only affected by the weather, but also by possible diseases and pests. Later, during and after harvest, further losses may occur due to processing, storage and transport techniques. These losses are particularly difficult to estimate given the extremely diverse factors that determine them.

The second and third stages of the supply chain, on the other hand, involve the processing and operations of the harvest and its subsequent transformation into edible food. In these stages, waste is due to food processing waste, partly physiological, partly due to the limits of the technology and processing used. The wrapping/packaging process with the choice of the most suitable materials for that type of food can also play an important role in waste prevention.

The fourth step takes into account the wholesale and retail distribution process of each food. Due to compliance with certain quantitative and aesthetic regulations and standards, marketing strategies and logistical requirements, most of the waste is food that is not sold.

The last two stages sleep related to the final consumption that usually takes place in restaurants and home kitchens. The waste that occurs at these two stages is mainly due to: the quantity of portions or food prepared, the purchase of too much food, the inability to consume by the use-by date and the difficulty of correctly interpreting the information provided on the label [4].

So far, the main reason for household leftovers is people's bad habits in choosing products when shopping, as large quantities of food are discarded due to the high standards of supermarket products.

Very often large quantities of fruit and vegetables are not selected for sale simply because they do not meet certain aesthetic qualities, while other products are generally not bought because they are considered to be of poor quality. This shows that much depends on consumers and individual purchasing and consumption choices. The act of discarding food is a voluntary act either because of food spoilage or because it is saturated due to forgetfulness or oversupply.

In addition to this, most food losses are intentional and are caused by inefficiencies in the food system, such as insufficient access to technology and energy, infrastructure and logistics, the market and management and capacity constraints on the part of supply chain participants.

The phenomenon of urbanisation, for example, has resulted in the progressive lengthening of the food supply chain to meet the food needs of the urban population. The greater distance between the place of production and the place where final consumption takes place creates the need to transport food over greater distances, with the need to improve transport, storage and sales infrastructure to avoid additional losses [5]. Other causes of food loss are climatic and environmental factors, the spread of disease, the presence of pests, the limitation of agricultural technology, economic reasons such as aesthetics, quality standards and market rules and last but not least consumer behaviour.

In general, the losses caused by these factors vary depending on the type of crop, season and production area. In addition, unfavourable weather events increase the loss of some crops especially at harvest time.

The climate factor is one of the dynamics that most influences the loss of food in the food system. Future changes and variability in weather conditions are expected to increase in frequency and intensity; this will lead to further increases in post-harvest losses. Extreme weather events, such as droughts or floods, can destroy all crops and livestock and consequently also the infrastructure of the supply chain. Irregular rainfall, on the other hand, can cause the loss of harvested agricultural products, damage the drying process and create wet conditions that develop pathogens.

When food is wasted, we must remember that it is economically harmful to us and to our planet, especially when we think of all the resources that are involved in producing it (Fig. 2.1).

To summarise, the main causes of food waste are as follows:

- Misinterpretation of words: in general, consumers do not read carefully what is written on product labels, e.g. "best before", which is understood to mean the quality of the food, and "use by", which relates to the safety of the food. In general, consumers are more willing to choose food with a longer "life", which leads to an increase in the number of unsold products and at the same time to an increase in the waste of food that could have been consumed.
- Bad supply management: this can lead to excessive food purchases.
- Poor food management: e.g. food does not smell or taste good, food is covered in mould, food has expired, food has been forgotten in the fridge, as a result of poor management in the portion preparation stage.
- Limited knowledge: of ways to minimise waste, e.g. reuse of leftovers.
- Lack of awareness: particularly of the economic and environmental impacts of the waste we produce.
- Household income: lower income households have less food waste than higher income households.
- Seasonality: this generally causes more waste in summer than in winter.

Clearly, the combination of these causes or bad practices has an extremely significant overall impact on the environment as well as on economic and social levels.

Environmental Impact: Completely unnecessary natural resources will be used to make food that will not be consumed and brought to the table, and at the same time, atmospheric emissions and waste accumulation will be produced.

Agro-food waste will also create waste in the use of fertilisers and pesticides.

To estimate the impact of food on the environment, it is necessary to consider its entire "life cycle", which covers all stages of the food chain.

Generally, there are three indicators considered to assess environmental impact: carbon footprint, used to estimate greenhouse gas emissions generated by processes; ecological footprint, used to estimate the impact of a given population's consumption on the environment: quantifying the total area of terrestrial and aquatic eco-systems required to sustainably provide all the resources used and absorb all the emissions produced; and water footprint, a specific indicator of freshwater use and is constructed to express both the quantities of water resources actually used and the way in which water is used.

Economic Impact: There are two main systems for increasing economic impact, namely production costs and market prices of raw materials.

In the first case, the value of a good is proportional to the resources needed to produce it. For this reason, the economic impact can be estimated as the "value lost through waste", and the cost incurred to obtain an individual good is used as the

It is estimated that up to **10%** of food waste is linked to **DATE LABELLING** on food products

CONSUMERS RECOGNISE THEIR ROLE IN PREVENTING FOOD WASTE

In Europe about **88M TONNES** of food are wasted annually

According to EU citizens the following actors have a role to play in preventing food waste:

Actor	Percentage
CONSUMERS	76%
SHOPS AND RETAILERS	62%
HOSPITALITY AND FOOD SERVICE SECTORS	62%
FOOD MANUFACTURERS	52%
PUBLIC AUTHORITIES	49%
FARMERS	30%
DON'T KNOW	2%

49% of Europeans think that **better** and **clearer information** on the meaning of **"best before"** and **"use by"** dates would help them **waste less food** at home

Flash Eurobarometer 425, Food Waste and Date Marking , October 2015 available here:
http://ec.europa.eu/food/safety/food_waste/eu_actions/date_marking/index_en.htm
Market study on date marking and other information provided on food labels and food waste
prevention. ICF in association with Anthesis, Brook Lyndhurst and WRAP, 2018.

Fig. 2.1 European Commission: food waste—communication materials. *Source* https://food.ec.europa.eu/system/files/2020-06/fw_eu_actions_date-marking_infographic_en.pdf

standard for calculation. In the second case, the value of the good does not depend on the cost of production, but on its utility, represented by the market price.

In fact, the economic impact of waste can be estimated using the "market prices of individual goods" as the standard of calculation; and an evaluation based on the theory of welfare economics can also be added to estimate food waste as an impact on society.

Thus, in the calculation, not only the market price must be considered, but also the negative external effects generated, so that the willingness of the company to pay for the environmental impact is added to the price. Furthermore, considering that most of the land is used in less useful ways than other alternatives, such as for the production of non-edible food, it can also be assessed by calculating the opportunity cost of the agricultural land used for food production.

Social Impact: On a social level, food waste, on the other hand, impacts on "food security", defined in 1996, internationally by the World Food Summit as: "All people, at all times, can obtain sufficient, social and economic material to sufficient, safe and nutritious food that meets their dietary needs and preferences to live an active and healthy life" [6].

Food security includes the national food supply, not only as an energy requirement, but also as an adequate supply of nutrients. Literature provides various estimates of the amount of daily energy a person needs for a balanced diet, but on average this amount is considered to be around 2700 kcal.

Although supply data show that the global food supply is sufficient to meet the energy needs of the population, there is evidence that there is a problem of malnutrition in the world.

These problems can be traced back to food supply difficulties due to high levels of poverty or conflicts in specific societies.

In fact, there is a close correlation between areas with a high percentage of under-nourished people and areas with a high percentage of extremely poor people, which shows how poverty can lead to an inability to produce or purchase the food necessary to maintain nutrition. Unlike the phenomenon of malnutrition, in a society where the food supply is sufficient and accessibility to food is guaranteed, food wastage will increase, even in the form of overeating [5–9].

Indeed, the number of people consuming more calories is increasing, exacerbating the problem of obesity.

In this sense, the problem of food waste is mainly caused by an economic factor, especially in developed countries, where the amount of food is high and "cheap" compared to the income of households, who therefore do not bother to avoid waste, and by a factor of market trends defined as "ripe for waste", i.e. the moment food is wasted, sales increase. Therefore, food waste becomes a global problem, found among both low-income and high-income countries.

According to FAO global estimates, about one third of the food produced worldwide for human consumption is lost or discarded each year, i.e. about 1.3 billion tonnes; if we take into account that inedible food, then the figure increases to 1.6 billion tonnes per year [10].

Furthermore, according to research conducted by Andrew F. Smith, along the entire supply chain, "from the field to the fork", on average, only 43% of the produce grown for food is actually consumed [7] (Fig. 2.2).

Fig. 2.2 Huge food waste in urban context (CC0)

Industrialised (high-income) and developing (low-income) countries waste about the same amount of food: 670 and 630 million tonnes respectively. Figures reported in the 2011 global studies (produced only ten years ago) showed the following.

Amounts lost and wasted per capita and per year

	Total (kg)	Production and sales (kg)	By consumers (kg)
Europa	280	190	90
North America and Oceania	295	185	110
Industrialised Asia	240	160	80
Sub-Saharan Africa	160	155	5
North/West Africa, Central Asia	215	180	35
South and Southeast Asia	125	110	15
Latin America	225	200	25

From the study in question, it could be noted that consumers in the richest countries squander practically the same amount of food (222 million tonnes) as the net food production of Sub-Saharan Africa (230 million tonnes): the loss and waste of food would also mean a significant squandering of resources (land, water, energy, labour, capital) that would not only entail a lack of rationality, hardly sustainable, but a negative impact associated, as has been pointed out, with the greenhouse effect and, therefore, with global warming and climate change [10].

There are big differences between developed and developing countries; starting with the preparation and sowing of the soil, and finally the cultivation or rational use of water, chemical fertilisers and pesticides; all leading to completely different yields, which are the main reasons for losses.

The wastage that occurs in low-income countries is centred more in the first stages of the food chain (field losses, first processing, field transport, storage). Thirty per cent of it is mainly caused by lack of technology, production tools and also poor food preservation, while 14% occurs due to the fault of consumers.

For high-income countries, on the other hand, 35% is caused by consumers and traders, in particular by the wrong habits of individuals, while the remaining 21% is caused by the preceding stages: production and distribution (processing and packaging, marketing, consumer wastage).

The US Department of Agriculture (USDA) estimated that 30% of all food produced in the USA is wasted every year, which is worth 48.3 billion dollars (32.5 billion euros) [8].

Half of the water used for its production is also squandered. Organic waste in the USA is the second-most abundant component in landfills, which in turn are the main source of methane gas emissions.

Even in harvesting, processing and storage, there are considerable differences between countries. In developing countries, losses are generally due to: labour-intensive and small-scale agriculture; technical skills; early harvesting; due to urgent need for food or lack of money; the usually inefficient harvesting method; inadequate infrastructure; lack of transport; food storage due to incorrect temperature; use of pesticides and, finally, lack of effective logistical organisation to ensure proper storage during transport. On the other hand, in higher income countries, where the best technology, infrastructural equipment, agronomic expertise, the most advanced technology and generally more favourable environmental conditions are available, the level of loss is much lower.

According to the latest data from the European Commission, food waste in EU countries in 2012 amounted to some 88 million tonnes per year (173 kg per person), which would represent about 20% of all food produced at a cost of 143 billion euros and some 170 million tonnes of CO_2 caused by the production and disposal of food waste [11].

In a few years, the data have increased dangerously, and according to the most recent data published by Eurostat, the waste generated in the years 2020, 2021 and 2022 amounted, in Europe, to an average of about 226 million tonnes, dumped (23%), incinerated (27%) or destined for recycling processes (30%) and composting (18%), generating between 14 and 15% of the world's food waste [12].

Given the dimensions assumed by the phenomenon of food waste and especially the extent of its impacts, the European Union (EU) has developed an assessment of food systems in terms of nutritional challenges, agriculture, food loss and food waste, generating guidance and actions to reduce them.

Specifically, in recent years, the EU has carried out three important measures:

(1)   The publication of guidelines on food donation, which helps to remove obstacles to food redistribution within the current European regulatory framework by empowering member states to choose the best solutions to address the problem at national level [13].
(2)   The updating of guidelines on the use as feed of food no longer intended for human consumption, which will facilitate the safe use as feed of former food products (in line with the food use hierarchy) and avoid food waste; [14]
(3)   The adoption, starting in 2020, of a new common methodology to measure food loss and waste (FLW), with the aim of pushing member states to flexibly set up a monitoring framework and provide new data on levels of food waste aimed at the publication of a first Pan-European report on the topic [15].

At the national level, noteworthy policy responses can be observed in France and Italy, which passed two laws sharing the idea that changing food distribution is an effective and coherent remedy against the unacceptability of waste [16].

In 2016, France was the first country in the world to enact a national law against food loss and waste, with specific obligations and penalties for all retail businesses whose surface area exceeds 400 m^2, which were obliged to sign an unsold food donation agreement with one or more charitable organisations [17].

Also in 2016, Italy enacted a new law (the Gadda law) [18] to facilitate food donations by streamlining the bureaucracy that hinders them, relaxing food safety and labelling requirements and offering tax incentives (i.e. deductions from waste taxes).

Overall, although the tools are providing more and more options to adopt more sustainable lifestyles and reduce the level of food waste, these changes have not yet reached the targets of the goals set in the 2030 Agenda, although there has been a general rise of good practices in the fight against food loss and waste (Fig. 2.3).

Poster raising awareness of bad food production and consumption practices, presented during the 'Festinar' event of the Creative Food Cycles project (Fig. 2.4).

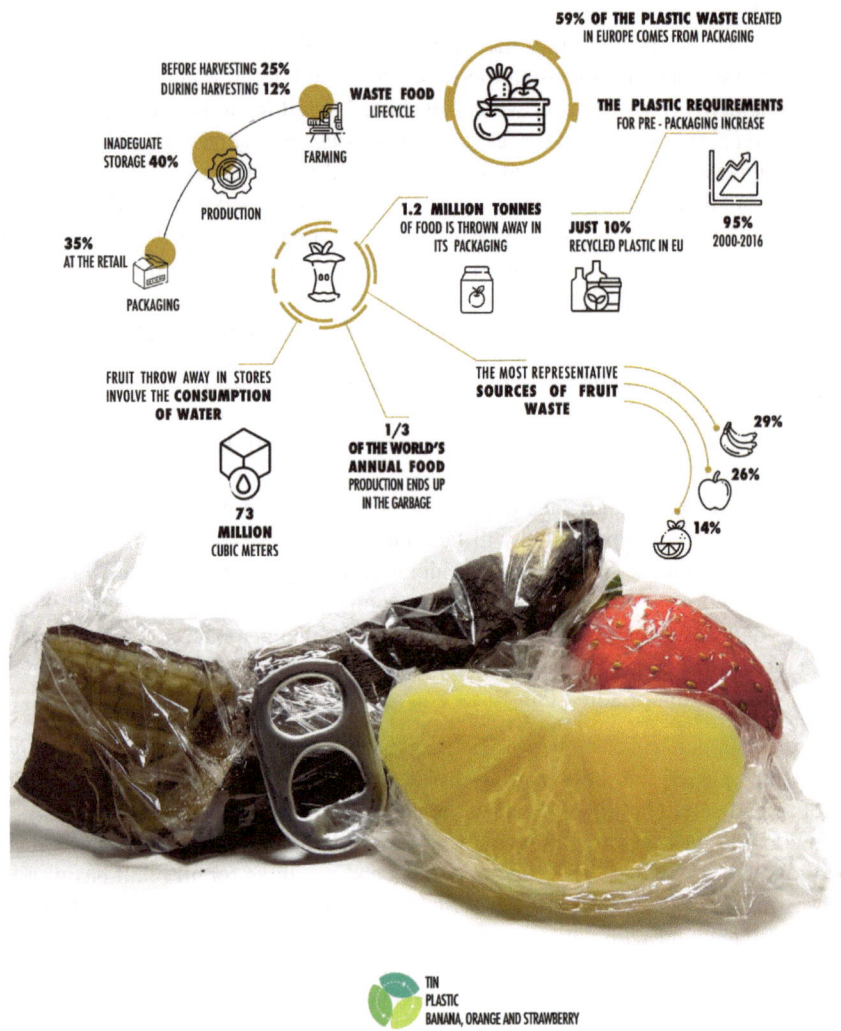

Fig. 2.3 Alessia Moi and Giorgia Parisi. Underrated beauty (GIC.Lab-UNIGE 2019)

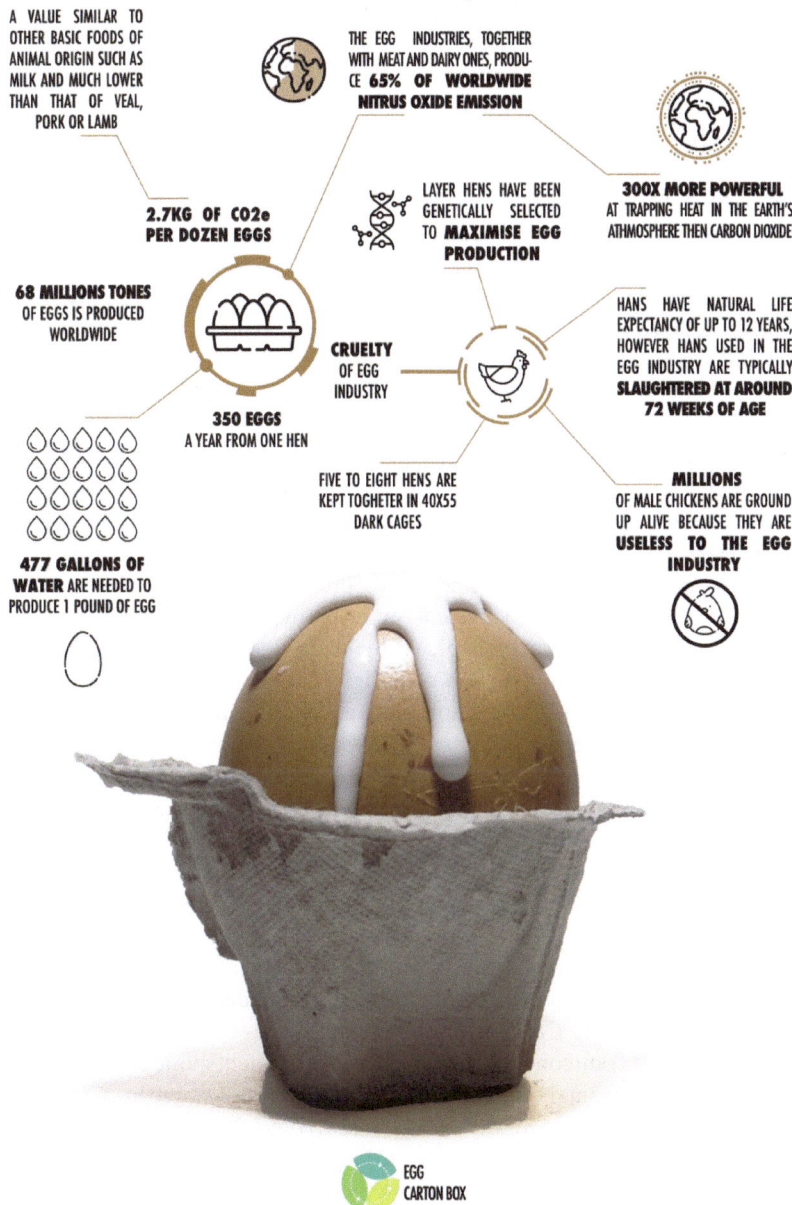

Fig. 2.4 Alessia Moi and Giorgia Parisi. Underrated beauty (GIC.Lab-UNIGE 2019)

Poster raising awareness of bad food production and consumption practices, presented during the 'Festinar' event of the Creative Food Cycles project.

Thanks to increased awareness of the problem through awareness-raising campaigns and increased media attention to the issue, many European cities are launching important initiatives to counter the paradoxes of the food supply chain and establish a true circular food economy. Working with production and consumption surpluses, but also with the waste associated with organic matter and packaging itself, therefore seems indispensable today to ensure a new type of alternative solutions that are more consistent with the conservation of our habitats. Obviously, one way to reduce food waste is to reduce spoilage by better planning purchases, improving nutrition, avoiding spontaneous or compulsive, unnecessary consumption and storing food properly (avoiding an excessive accumulation of perishable stocks).

Another possible response is the so-called "smart packaging", not only aimed at providing accurate information on food, its origin and expiration dates, etc., but also through the simple application—on the packaging itself—of interactive codes linked to apps. Informative apps that can be linked to functional or sensitising platforms or tutorials, etc. The use of new types of packaging with, for example, temperature-sensitive inks, plastics or gels that change colour over time (or when exposed to harmful oxygen) can be combined with the use, for packaging, of biomaterials (bioplastics, bio-celluloses, bio-textiles) derived from food waste itself, helping to close the usual vicious circles.

Food, in any case, can be composted to produce soil and fertilisers, to feed animals and can be used to produce energy or biofuels; but it can also generate—with ever greater scientific-technological efficiency—new biomaterials and derivatives, destined to favour the creation of designs and processes that end up being, in turn, bio-degradable, increasing soil cultivation capacities. These new ways of bio-production speak, then, of a *Second Life*—or *Second Role*—for food, beyond its strict food definition [19].

In this sense, the management and treatment of organic waste (or bio-waste) and its possible reuse or recycling is one of the great global challenges, both for producing companies and for public and private decision-makers and managers and other actors involved in the value chain. The agricultural sector is one of the major players involved, and the possibility of promoting a second life for food waste and surpluses (*Second- Life-Food-Waste* or simply *Second-Life-Food*) is obvious and evidently linked with the paradigmatic premises of circular processes issued from the very known *Cradle to Cradle* approaches [20].

Europe has set common goals in waste management policy to try to limit the growing environmental and economic problems caused by bad practices. For example, the new *Waste Framework Directive (Directive 2018/851)* sets as one of its priorities the separate collection of bio-waste by the end of 2023 in all Member States and the implementation of new targets for the re-use, recovery and/or recycling of waste, to reach 65% by 2035.

In this sense, more and more techno-scientific and eco-business proposals tend to generate new opportunities to transform waste and by-products into elements of high added value through sustainable, low environmental impact and highly efficient processes. These "revalued wastes" can be categorised into three main groups:

- Sludge from wastewater treatment plants: obtained during wastewater treatment processes and from which bio-celluloses can be extracted (approximately 40–50% of their composition).
- Lignocellulosic wastes: including agro-forestry wastes and some industrial and urban wastes with a fibrous/woody profile.
- Agro-food waste: food waste from the food industry (including beverages and livestock) and from the food packaging and wrapping industry. The latter is the main focus of the following pages.

2.2 Beyond Food Waste: Bio-Agro-Techno Materials

We have already pointed out how throwing away food—and products associated with food—has become a daily occurrence in about eight out of ten households around us.

Figures for 2018 show that around 1339 million kilos/litres of food were thrown away, an unsustainable waste of food to which various types of packaging, not always biodegradable, are added. This wrong dynamic must be addressed as a priority, not only with new domestic habits, but also with new alternative recycling systems. During the last two years, more than a third of all food produced in the world (2.5 billion tonnes) has been lost or wasted each year. We are talking about 30% of fundamental resources. This number is even more surprising, given the large number of hungry people in the world. Wasted food is not only a wastefulness but a social injustice.

However, in recent years, a new awareness and sensibility has developed among consumers who are increasingly conscious of the socio-environmental values of products, the origin of raw materials and production processes.

In fact, as shown by the Two Sides Europe research [22], presented in April 2020, the profile of the new consumers emerging in Europe is increasingly informed and aware of buying more eco-friendly products. The packaging industry, for example, plays an extremely important role in purchasing decisions as packaging tends to be discarded once the product has been purchased, so this process focuses consumers' attention on the environmental impact of different packaging materials. Data confirm a growing trend in the preference to purchase products with packaging made of biodegradable, recyclable or reusable materials, such as paper and cardboard chosen for its compostability (72%) or glass chosen for its ability to be reusable (55%).

Seventy percentage of users say they take active steps to reduce the use of non-sustainable packaging, while 48% avoid buying from retailers who do not actively work to limit non-recyclable packaging, and 58% say it would be fair to apply specific taxation to discourage the use of environmentally impactful packaging.

As Jonatham Tame, Managing Director of Two Sides Europe, states: "Packaging is receiving more attention than ever in a bid to achieving a circular economy.

Consumers are becoming more aware of the packaging choices for the items they buy, which in turn is applying pressure on businesses—particularly in retail. The culture of 'make, use, dispose' is slowly changing".

Consumer awareness and sensitivity to the impact of packaging on our planet are, in fact, increasing. Sustainability has played a significant role in government policies around the world, gaining wide coverage in the media, public opinion and political debate. This process has certainly contributed to "educating" the growing number of consumers to develop an ecological consciousness (Fig. 2.5).

This process has certainly contributed to "educating" the growing number of consumers to develop an ecological consciousness.

This new awareness is consequently forcing industry, in its various sectors, to constantly evolve to meet consumer demands, as well as government policies to act on reducing environmental impacts in a concrete way.

The context of action today clearly defines a mission that can no longer be postponed: industry has the opportunity and the duty to support the modern sensitised, informed and active consumer by adopting a new sustainable approach aimed at the "zero waste" strategy. It is necessary to start by rethinking materials, which often consist of several components or combined materials that are difficult to separate or cannot be recycled. So we need to change the way we consume, rethink how products are sold and enhance the supply chain.

According to the *Circularity World Gap 2019* report [23] presented at the Davos Forum, only 9% of the 92 billion tonnes of raw materials consumed in the world are recovered and reintroduced into the system, following the principles of the circular economy. The gap to be recovered is still enormous and the urgency to do so is serious.

However, this need is fuelling a new creativity: those of already well-established companies, or innovative start-ups, which see the opportunity to save the Planet and offer original and sustainable products together with the recovery of waste materials.

This need has favoured a strong push towards design solutions for waste enhancement, thanks above all to the advancement of technological innovation in research and experimentation increasingly oriented towards environmental sustainability. Due to the number and quality of emerging projects, this decade could mark a crucial stage in the development of food waste recycling processes [24].

In this sense, there is a growing number of companies and start-ups that are transforming waste into a new resource, adopting the principles of the circular economy: a strategy implemented by several governments to guide the production system to pursue environmental improvement actions towards real sustainability goals. In a short time combine ethics and social responsibility with the legitimate desire of investors to make profits could become reality, as the utopian materials derived from food waste have become real.

One possible path is "bio-design" based on the use of new bio-organic (or simply "bio") materials generated from wasted food and packaging, reusing resources already available and discarded in the garbage, to turn them into new materials and products with a second life.

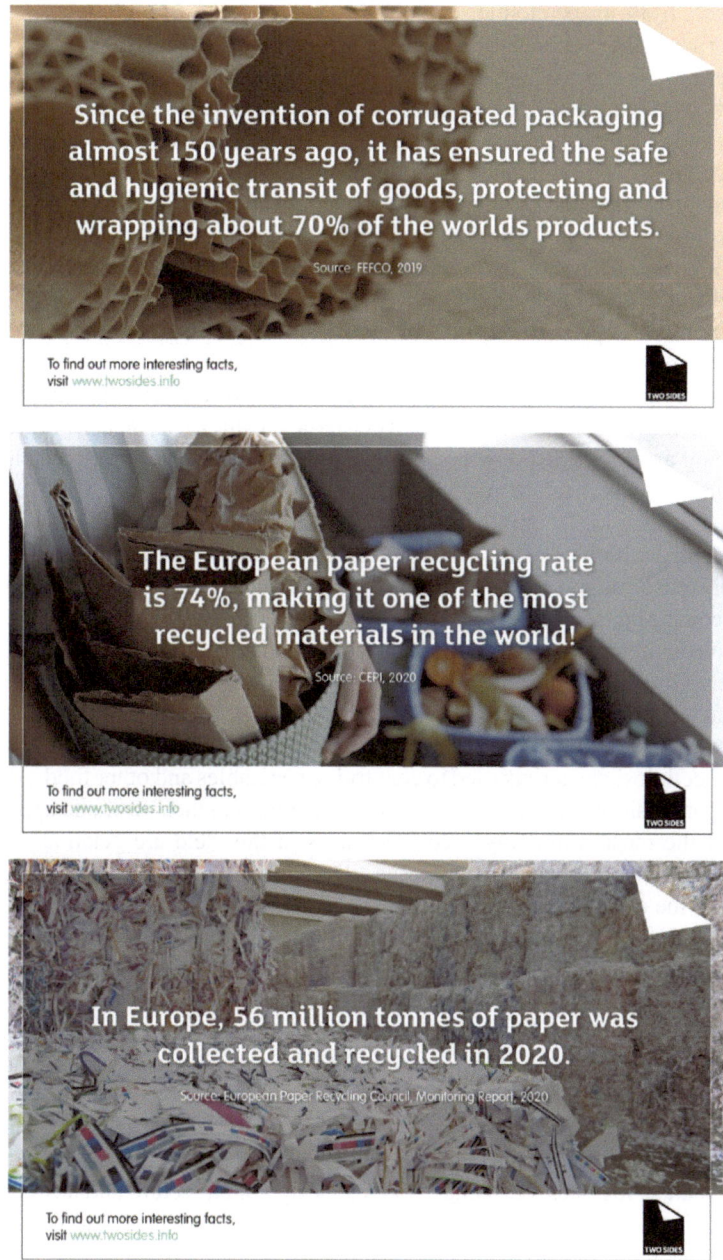

Fig. 2.5 Factographics from Two Sides Research, 2022. https://www.twosides.info

In fact, from an economic, ecological, technological and digital perspective, biomaterials and bio-products, as well as the positive management of organic waste, are emerging as some of the main protagonists of an increasingly eco-responsible and techno-advanced design, sensitive and innovative at the same time.

Fruit and vegetable remains, coffee grounds, crustacean shells, egg or cocoa shells, potato peelings, vegetable leaves and oils, milk serums, mushroom mycelia can be authentic raw materials for new bio-products translated into "other types" of skins, fabrics, plastics, rubbers, ceramics, cements, celluloses, dyes, varnishes, etc., beyond the usual synthetic products, but also beyond the traditional applications of food products in the cosmetics, alcohol, beverages or fertilisers' industries (Fig. 2.6).

The number of possibilities offered by waste is becoming remarkable thanks, in large part, to the increase in biotechnological knowledge.

The food industry itself is trying more and more to reconcile the needs of food safety with those of the new sustainability criteria, the market is, in fact, progressively developing and proposing new systems and alternative materials for containers and packaging that can meet the requirements of the new bio-industry.

"Smart packaging", "edible films" or "bioplastics" are some of the new products experimented with increasing qualitative precision and functional efficiency.

In this sense, the so-called "smart packaging" is packaging systems which, in addition to containing and preserving food, protecting it from the outside, play an additional role in maintaining (and even improving) food and, at the same time, providing interactive information on the quality of the food and its packaging, increasing its shelf life and shelf life.

More closely linked to food recycling, edible films are thin layers of edible polymeric or cellulosic materials used to coat fruits, vegetables and other foods, delaying their decomposition and improving both their colour and their aesthetic appearance. Some of the most commonly used substances in this field are gelatines, sucrose, cereal glutens, corn proteins or chitosan (derived from chitin, a structural element present in the exoskeleton of crustaceans).

In fact, bioplastics or plant-based plastics are natural polymers that are 100% biodegradable, produced from natural elements (fruits, vegetables, plants, but also dairy and fungal, etc.), generally during bacterial, biosynthetic or fermentation processes. Their growing development is evidence of the need to find alternative biomaterials to plastic, but with similar characteristics and from non-polluting sources.

This kind of experimentation on new biomaterials, so decisive in the food industry, is gradually taking on an increasingly important and integral role also in parallel sectors such as logistics, cosmetics, textiles, design, architecture and technology, as well as in the field of modern medicine, providing healthcare professionals with new perspectives in the development of more compatible organic elements or biological systems.

Innovation in the field of bone regeneration was one of the first pioneering fields in this type of application. The development of bioactive polymer-ceramic composites for the fixation of implants and the restoration of bone defects is now common and allows for increased strength and bio-integration of implants [25, 26].

Fig. 2.6 AI-generated bio-packaging (biopackaging, smart packaging, bioplastic, edible film, plant-based)

Bioengineers are already experimenting with the potential of food, such as the ability of amino acids in eggs to repair damaged organs. Egg whites contain large amounts of amino acids needed to produce collagen, a biological macromolecule that makes up 35% of the human body's protein and is definitely effective for the regeneration of human organs. Various methods to formulate vascular bio-inks (using egg white) can also be bio-printed into functional tissues. Three-dimensional bioprinting is very promising for the development of specific tissues and organs and for alleviating the shortage of organ donors.

By-products of the aquaculture and fishing industry, such as discarded scales, organs and skin, also offer new perspectives in the experimentation of collagen-based skin grafts as an alternative to skin grafts to repair wounds and injuries.

These thin 2D collagen bio-membranes strongly reduce water evaporation and protein diffusion and act as a barrier against bacterial infiltration, protecting against more than 99% of the tested bacterial species, thus functionally mimicking the epidermal layer (Fig. 2.7).

In this sense, the generation of new biomaterials offers today promising results in the current R + D + D innovative processes and generates new paradigms able to find solutions to the invasive food-waste problems and alternatives to the unsustainable synthetic materials. The possibilities to generate biomaterials 100% biodegradable, compostable and even edible, inspire new proactive approaches—beyond old nihilistic topics—and sustainable premises based on circular processes.

The aim of increasing public awareness to establish ethical and eco-durable systemic values in our ways of life calls to a new propositional research carried

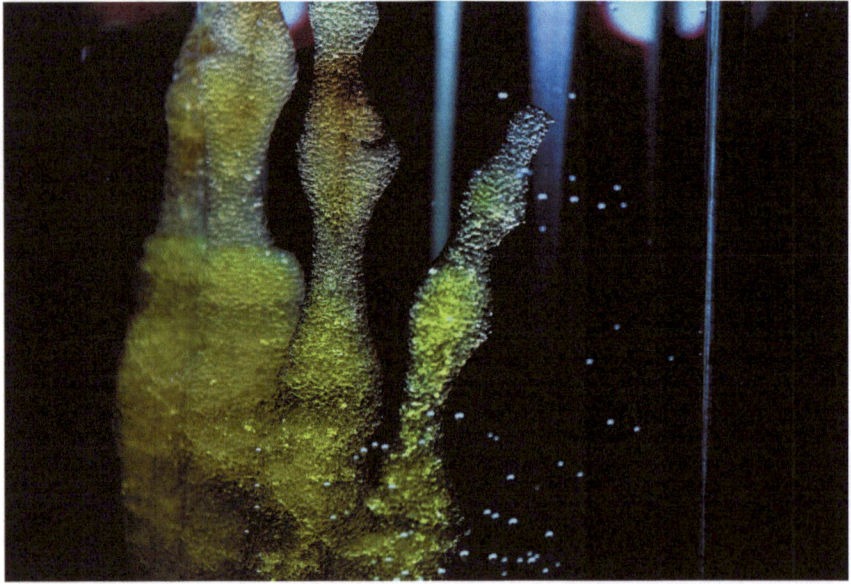

Fig. 2.7 Stem cell image. Regenerative reliquary. *Author* Amy Karle, 2016 (CC)

out with decision and precision—in terms of ingredients, characteristics, times and formulations—capable of creating biomaterials or bioplastics with sufficient organic matter or food waste contents but with optimal physical properties for their possible concrete applications in various sectors, contexts and environments.

To obtain these biomaterials, food waste (generated from fruit and vegetables, cereals, eggs and milk, coffee grounds, seafood) is used as a key factor, with the addition of organic materials and thickeners (maize starch, vinegar, vegetable glycerine, potato starch, agar–agar, bio-binding agents), as we will see in the following chapters.

This new mix of research (scientific, creative, technological and environmental, but also ethical and aesthetic) can be highly sophisticated when generated in centres with important infrastructures and facilities, or simply domestic, when made with basic kitchen tools.

A huge number of design studios involved in the topic are now developing "biomaterial kits" aimed at the creation of bio-products through the use of basic ingredients, to enable users to generate their own biomaterials or bioplastics at home (by adding only water and treated organic waste). These products aim to raise consciousness of the possibilities associated to food waste treatments and biomaterials.

The fact that the new materials can be created "at home" favours a change in the mentalities, opening new behaviour (and neighbour) dynamics. Social networks linked with informational tutorials or digital apps., stimulate sharing processes where every contribution can be exchange in "open source" platforms, helping to increase a collective knowledge and generating new "Engaged-Communities" where every member help each other with the sole purpose of exploring, learning and moving forward more sustainable habitats.

Evidently, there can be many applications (papers, household items, food bags, new packaging elements) depending on the characteristics of the generated biomaterials that can withstand organic biodegradation over time (even years).

2.3 The Chemical Factor: From Additive to Multiplicative Elements

As anticipated, the desire to promote a new WZ ("Waste Zero") circular economy calls for the promotion of experimental processes associated with waste products from various fields and transformed into real raw materials for new products.

This redefinition takes place mainly through biological, mechanical and, of course, chemical processes [27]. Chemistry has been central to the history of food production by influencing its production, preservation, protection and consumption, from the use of fertilisers, pesticides and preservatives, to the use of additives and artificial substitutes.

The very act of cooking, transforming the various combined qualities of ingredients to form new textures and flavours, is itself a chemical process [28, 29].

Already in ancient China, kerosene wax was burned to ripen fruit; the Egyptians coloured food with saffron and the Romans added alum (aluminium potassium sulphate) to bread to make it whiter, all of which are examples of the primitive use of food additives. In fact, the first food additive was probably salt, a valuable commodity for preserving foods such as fish and meat by dehydrating food to limit bacterial growth [30].

The proto-industrial revolution and the wars and conflicts of the late-eighteenth and early-nineteenth centuries helped to promote new preservation and transport technologies for initially military or logistical uses (the *can* was to become a basic packaging technology that was able to combine the sealing of food in an airtight container with heat sterilisation).

The use of food additives was also to increase exponentially during the industrial revolution, with (often toxic) compounds used extensively in food factories (whether to modify or enhance the colouration, texture, taste or flavour of products).

Concern about the toxicity (and carcinogenic effects) of additives would intensify, however, in the mid-twentieth century, when analytical chemistry made it possible to detect and measure the specific effects of additives.

As the food processing industry grew and chemistry became capable not only of synthesising new additives (thickeners, emulsifiers, sweeteners, colourants and, ultimately, artificial foods such as margarine, saccharin, etc.), regulatory organisms and chemistry protocols were being set up—almost in parallel—to control the quality and healthiness of products intended for human (and, to a lesser extent, animal) consumption. The growing concern about the use of chemical additives or artificial foods in feeding offerings—and the current socio-cultural trends, more eco-sensitive—have led consumers to seek more natural organic products.

In this regard, biochemistry is helping to understand how additives can have an adverse impact on our gut flora, which has given rise to the burgeoning probiotics industry. Thus, new bio-chemical and bio-medical studies, as well as socio-governmental regulations, are helping to variate our eating habits [31].

New technologies combined with new multidisciplinary research have helped to promote a more holistic and precise management of processes that are substantially more interactive and complex than the old unidirectional models would suggest.

In this sense, biotechnology and industrial bio-design—as well as biochemistry—are increasingly applied to new processes aimed at creating products (generally organic-based) that are easily degradable and with more sustainable production processes, capable of consuming less energy and generating less waste.

From the generation of the old "additive" elements, we have moved in a few years to a new concept based on "multiplicative" processes, capable of multiplying the organic properties of foodstuffs, but also their use and final destinations as material products, not only for *foods* but also for other polyvalent *goods* (with other valences and potencies) (Fig. 2.8).

Fig. 2.8 Kaiku-Living Color Project. Waste food Bio-alternatives to the use of artificial dying pigments (Nicole Stjernswärd, London, UK 2019). https://www.stjernsward.co/kaiku-living-color

2.4 Biomaterials, Food and Contemporary Innovation

We have already pointed out how the equation "BIO-technology + AGRO-culture + SOS-tenability + HYPER-materials" could help (from informational logics related to control parameters and intelligent management associated to precise indicators) to increase in a process-positive way not only the cultivated productions (preserving areas of great ecological value and reducing the use and consumption of soil and resources) but also to diversify the use(s) of agricultural productions for other destinations, reducing post-harvest losses, redefining or recycling food wastes (and their derivatives), diversifying the use(s) of agricultural production for other destinations or functions, reducing post-harvest losses, redefining or recycling food waste (and its by-products) [32].

One of the most recent applications of the combination of agriculture, biology, biochemistry and new technologies would be that of biomaterials, synthesised from natural materials or food or agricultural waste, using methodologies based on bio-efficient systems called to replace, for example, plastics and other petroleum-derived materials.

This need is fuelling a new creativity: those of already well-established companies, or innovative start-ups, which see the opportunity to save the Planet and offer original and sustainable products together with the recovery of waste materials. In recent years, the awareness of industries and consumers, increasingly attentive to the socio-environmental values of products, to the origin of raw materials and production processes, has favoured a strong push towards design solutions for waste enhancement, thanks above all to the advancement of technological innovation in research and experimentation increasingly oriented towards environmental sustainability.

As mentioned earlier, this decade marks a turning point in the development of food waste recycling processes given the huge growth of companies and start-ups that are turning waste into a new resource, embracing the principles of the Circular Economy and opening up new market opportunities [24].

Food waste becomes the primary raw material in the production of new biocompatible and sustainable materials such as: new biomaterials for construction, new textiles for the fashion industry, new bio-products for the design world and new bio-packaging for food. Among the new innovative materials resulting from experimentations on the second life of agro-food waste, we have:

- **Bioplastics**, derived from artichokes, sugar beets, shrimp shells, cactus, but also from potatoes, frying oil, grains, algae, etc.

  Curiously, bioplastics (PHA) were discovered by the French chemist Maurice Lemoigne in 1925, even before Staudinger formulated his theory of polymers and the development of synthesis methods for early plastics (LDPE, PVC, PS) around 1935. The research and the modern green chemical industry are focusing on the production of PHA due to the versatility of this biodegradable molecule. The data released by the European Commission with the report "A European Strategy for Plastics in a Circular Economy" inform us that every year plastic waste amounts to 25.8 tonnes, 31% of which is landfilled. One of the most critical issues in terms of environmental sustainability is the very short life cycle of plastic, a dissipated value that fluctuates, according to estimates, between 70 and 105 billion per year [33].

  The world has produced 8 billion tonnes of plastic since the 1950s and demand is still increasing. "But we can't go on using fossil fuel-based materials. About 6–7% of every barrel of oil is used to make plastics" said Paul Mines (CEO of Biome Technologies, UK), which has spent £ 5m in the last five years on bioplastics research. Using plant materials is feasible, said Prof Simon McQueen-Mason, at the University of York, replacing half of the nation's plastic bottles could be done using just 3% of the sugar beet crop, 5% of wheat straw or 2.5% of food waste.

  The transformation of horticultural waste into bioplastics, sourced entirely from sustainable sources, is a perfect example of a circular economy, producing a new environmentally friendly and compostable material that drastically reduces the cost of organic waste disposal, while remaining connected to the territory to which it returns at the end of its life cycle. To date, the negative aspect is the cost. Producing plastic from fossil sources is an extremely cheap process.

  So much so that in the world, every year, 250 million tonnes of conventional plastic are used, while the organic one remains below 1%.

  Today, the real revolution lies in developing production processes that reduce production costs and encourage the total replacement of the traditional plastic with the biodegradable one (Fig. 2.9).

Fig. 2.9 Bioplastic from algae developed by Atelier Luma, France, 2017. The project Jellyfish Farm demonstrates how bioplastic derived from algae can be mixed with starch and could replace traditional plastics, creating a versatile biodegradable material. www.luma.org/en/arles.html

- **Bio-ceramics and bio-mortars**, understood beyond their traditional connotations. Current research on food waste, focusing in particular on the properties of calcium carbonate (from egg shells, mollusc shells and others) or the particular versatility and capacity of coffee sediments, allows, through circular approaches, to experiment with these elements, converting them into raw materials for bio-composites.

  By means of traditional and/or more sophisticated production processes—such as 3D printing or with robotic arms—food waste, mixed with biopolymers, is transformed into bio-ceramics and/or malleable mortars, which give rise to modelled, folded, cut or printed products.

  New biomaterials adaptable to design or construction with varied geometries that can be used for a large number of purposes, especially in the home (crockery, cutlery, glasses, cups, jugs...) or in construction (bricks, absorbent surfaces, building elements...). Thanks to their low viscosity, bio-ceramics, bio-cements and bio-mortars are resistant and suitable for moulding (Fig. 2.10).

- **Eco-textiles**, derived from oranges, pineapples, soy, corn, crustaceans, grapes, milk, from apples and natural dyes derived, instead, from white artichoke, coppery onions, pomegranate peel and cherry and olive tree pruning residues. Considering that world production of garments is destined to grow by 63% by 2030, the

Fig. 2.10 Andrea Montaldo. DishBratta (GIC.Lab-UNIGE, 2019) Environmentally sustainable tableware made from coffee grounds and natural thickeners, exhibited at the Creative Food Cycles Workshop event on new biomaterials made from food waste

potential of an ecological textile supply chain is enormous, up to representing 20% of the sector's turnover in Italy (currently 4.2 billion).

Nowadays, a T-shirt uses an average of 2700 litres of water, generates high levels of CO_2 emissions and is made mainly from dyes and synthetic fibres.

Natural dyes related to the use of plant and animal fibres, from wool to silk, from linen to hemp, can also be a valid aid to the increasing problems of allergic contact dermatitis due to synthetic dyes. By recovering plants and cultivation waste for dyeing, it helps to redevelop abandoned or degraded areas and to consolidate territories while protecting biodiversity and landscape (Fig. 2.11).

- **Green paper**, derived from banana, oranges, apples, grapes and hazelnut shells. The residues of citrus fruits, grapes, cherries, lavender, corn, olives, coffee, kiwi, lentils, beans and almonds are the main natural raw materials that, saved from the dump, are used for the production of these new papers. However, the greatest amount of paper and cardboard remains in packaging, which has become an important factor for our economy. The demand is constantly increasing, driven, e.g. from purchases and shipments to online e-commerce, but at the same time, functional requirements become increasingly complex, both in the field of protection, information display, product encoding or convenience. The realisation of more sustainable packages is an important task to follow these growing trends.

Fig. 2.11 Eco-textile from waste of fruit and vegetables developed by two designers—Koen Meerkerk and Hugo de Boon, The Netherlands, 2021. The project Fruit Leather seeks to create a variable and versatile material that can be turned into footwear, fashion accessories, upholstery and furniture. Fruit leather with permission from K. Meerkerk and H. de Boon, https://fruitleather.nl

Consumer demands and environmental needs are the main drivers for developing sustainable packaging, and increasing the use of recycled material is one of the key approaches to a sustainable economy. The vegetable by-products previously treated as waste become a functional part of the packaging for the final food products. Conventional wood-based fibres are replaced, protecting natural resources and increasing the proportion of recycled material in the carton. Thus, the food parts of the plant meet again the inedible parts, adding economic and ecological value to the entire production chain and to the consumer's final product (Fig. 2.12).

According to a Eurostat study entitled "Environmental economy statistics on employment and growth", in the last 15 years, the wealth produced by the green economy in EU member countries has increased from 135 to 290 billion euro, with an impact on the product gross domestic product (GDP) which, over the same period, grew from 1.4 to 2.1%. In terms of turnover, according to the Eurostat surveys, the green economy has grown exponentially, reaching a turnover of 700 billion euros, while at the employment level, green jobs have now gone from 1.4 million to 4.1 million people across Europe [10–12, 19–21, 24, 27–37].

Fig. 2.12 Green paper from waste of fruits and vegetables developed by Unige designers within the Creative Food Cycles project. See 3.3 and 3.4 for more information

Not only the European Union, but at a global level, the Sustainable Development Goals—Agenda 2030 (SDGs) of the United Nations promote and face the global challenge of the green economy, a challenge that can no longer be delayed given the unsustainability of the current model, in key to fighting climate change and decarbonising the system, with the aim of guiding a transition to sustainability that includes the transition from a linear economy to a circular economy, the correction of imbalances in our food system, the energy of the future, buildings and mobility [24, 32, 36, 37].

Biomaterials today represent a concrete solution to counter these issues, rethinking (and even improving) existing products, with the aim of generalising the use of alternative systems. To paraphrase Camila Castro Grinstein of Studio Etimo: sowing design, cultivating technology and harvesting bio-manufactured products allows us to announce a future that is "planted" as a timely and possible challenge (Fig. 2.13).

Fig. 2.13 Bio-leather, made from discarded vegetables and natural dyes with permission from Estudio Etimo Lab—Camila Castro/Buenos Aires, 2017. www.etimobiomateriales.com

References

1. FAO (1981) Food loss prevention in perishable crops. https://www.fao.org/3/s8620e/s8620e00. htm
2. FAO (2011) Global food losses and food waste. Extent, causes and prevention. https://www. fao.org/3/i2697e/i2697e.pdf
3. Parfitt J, Barthel M, Macnaughton S (2010) Food waste within food supply chains: quantification and potential for change to 2050. Philos Trans R Soc B: Biol Sci 365:3065–81
4. Fondazione Barilla (2021) L'Europa e il cibo. Garantire benefici sull'ambiente, sulla salute e sulla società per la transizione globale. https://www.fondazionebarilla.com/wp-content/upl oads/2022/05/LEuropa-e-il-Cibo.pdf
5. Centre for Non-Traditional Security Studies (2011) Mind the gap: reducing waste and losses in the food supply chain
6. FAO (1996) The state of food and agriculture. https://www.fao.org/3/w1358e/w1358e00.htm
7. Lundqvist J, de Fraiture C, Molden D (2008) Saving water: from field to fork—curbing losses and wastage in the food chain. In: SIWI Policy Brief. SIWI
8. Nellemann C et al (2009) The environmental food crisis—the environment's role in averting future food crises. UNEP
9. Falkenmark M, Rockström J (2004) Balancing water for humans and nature: the new approach in ecohydrology. Earthscan, London
10. FAO (2011) Global food losses and food waste—extent, causes and prevention. Rome. https:// www.fao.org/3/mb060e/mb060e.pdf
11. Oreopoulou V, Russ W (2007) Utilization of by-products and treatment of waste in the food industry. Springer, New York. https://books.google.co.uk/books?id=9G1M9Z8qgJ4C& printsec=frontcover&source=gbs_v2_summary_r&cad=0#v=onepage&q&f=false

12. Eurostat (2020) Energy, transport and environment statistics report. https://ec.europa.eu/eur ostat/documents/3217494/11478276/KS-DK-20-001-EN-N.pdf/06ddaf8d-1745-76b5-838e-013524781340?t=1605526083000
13. European Commission (2017) Communication from the Commission of 16.10.2017: EU guidelines on food donations (No 61). https://food.ec.europa.eu/safety/food-waste/eu-actions-aga inst-food-waste/food-donation_en
14. European Commission (2018) Communication from the commission. Guidelines for the use as feed of food no longer intended for human consumption (2018/C 133/02). https://eur-lex. europa.eu/legal-content/EN/TXT/PDF/?uri=CELEX:52018XC0416(01)&from=EN
15. European Union (2019) Commission Delegated Decision (EU) 2019/1597 of 3 May 2019 supplementing Directive 2008/98/EC of the European Parliament and of the Council as regards a common methodology and minimum quality requirements for the uniform measurement of food waste levels. https://faolex.fao.org/docs/pdf/eur189612.pdf
16. Ferrando T, Mansuy J (2018) The European action against food loss and waste: co-regulation and collisions on the way to the sustainable development goals. Yearb Eur Law 37:424–454. https://doi.org/10.1093/yel/yey015/5163090
17. French Republic (2016) Loi n° 2016–138 du 11 février 2016 relative à la lutte contre le gaspillage alimentaire. https://www.legifrance.gouv.fr/eli/loi/2016/2/11/AGRX1531165L/jo/texte
18. Republic of Italy (2016) Law No 166 of 19 August 2016, Dispositions concerning the donation and distribution of food and pharmaceutical products for purposes of social solidarity and for the limitation of waste. https://www.gazzettaufficiale.it/eli/id/2016/08/30/16G00179/sg
19. Stuart T (2009) Waste: uncovering the global food scandal: the true cost of what the global food industry throws away. Penguin, London
20. McDonough W, Braungart M (2002) Cradle to cradle: remaking the way we make things. North Point Press, New York (Farrar, Straus & Giroux)
21. De la Cruz M, Albaladejo P (2022) Your waste has a second life. https://www.infopack.es/en/new/your-waste-has-a-second-life
22. Two Sides, European Packaging Perefernces (2020) A European study of consumer preferences, perceptions and attitudes towards packaging. https://www.twosides.info/documents/research/2020/packaging/European-Packaging-Preferences-2020_EN.pdf
23. Circularity world gap 2019 report. www.circularity-gap.world
24. Julian P, Barthel M, Macnaughton S (2010) Food waste within food supply chains: quantification and potential for change to 2050. Philos Trans R Soc Lond B: Biol Sci 365(1554):3065–3081. https://doi.org/10.1098/rstb.2010.0126
25. Magee C (2020) The use of food waste in the development of biomaterials. https://www.inn ovationnewsnetwork.com/the-use-of-food-waste-in-the-development-of-biomaterials/6697/
26. MDPI Journals (2020, 2021, 2022, 2023) Materials collection, 3D printing in medicine and biomedical engineering. https://www.mdpi.com/journal/materials/topical_collections/3D_Print_Medicine_Biomedical_Engineer
27. Robison D (2013) OSU turns winemaking waste into food supplements and flowerpots. https://agsci.oregonstate.edu/story/osu-turns-winemaking-waste-food-supplements-and-flowerpots
28. Levenstein H (2012) Fear of food: a history of why we worry about what we eat. The Chicago University Press, Chicago
29. Kiple KF, Ornelas KC (ed) (2000) The Cambridge world history of food. The British journal of nutrition, vol 85, 6. Cambridge University Press, Cambridge. https://doi.org/10.1079/BJN 2001354
30. Smith AF (2015) Sugar: a global history. The Chicago University Press, Chicago
31. Shephard S (2006) Pickled, potted and canned: how the art and science of food preserving changed the world. Simon & Schuster, New York
32. UN DESA (2019) SDG good practices. A compilation of success stories and lessons learned in SDG implementation. https://sdgs.un.org/sites/default/files/2020-11/SDG%20Good%20Prac tices%20Publication%202020.pdf

33. A European strategy for plastics in a circular economy report. www.ec.europa.eu/environment/circular-economy/pdf/plastics-strategy-brochure.pdf
34. Eurostat, environmental economy statistics on employment and growth report. www.ec.europa.eu/eurostat/statistics-explained/index.php?title=Environmental_economy_-_statistics_on_employment_and_growth
35. Pitanti M (2021) ODE TO WINE. Ancient and contemporary values of the wine cycle. In: Pericu et al (eds) Creative food cycles experience. Goa CFC-festinar: a virtual banquet for an innovating research celebration. ADDDoc Logos, Genova, pp 99–110
36. UN (2022) Third global conference on strengthening synergies between the Paris agreement and the 2030 agenda for sustainable development, conference report. https://sdgs.un.org/sites/default/files/2023-03/the_third_global_conference_report_11.08.2022.pdf
37. IEA, IRENA, UNSD, World Bank, WHO (2022) Tracking SDG 7: the energy progress report. World Bank, Washington DC. https://un-energy.org/wp-content/uploads/2022/06/sdg7-report2022-full_report.pdf

Chapter 3
Food Second-Life as Innovative Research

3.1 Bio-experimental Researches and Technologies

As we have pointed out, more than 1.3 billion pieces of food waste end up in landfills every year. When food waste begins to decompose, it releases methane gas into the environment, contributing to climate change. CO_2 emissions are often considered the biggest contributor to global warming; however, methane traps 30 times more heat in the atmosphere. Various organisations, innovative industries, conscious companies and design companies are working to reduce the amount of food waste in the environment by supporting the development of new biomaterials.

It is well known that so-called biomaterials already play an important and integral role in modern medicine, providing healthcare professionals with essential technologies related to more compatible organic elements or biological systems.

Beyond the healthcare field, the generation of new biomaterials now also offers promising results in current innovative processes and generates new paradigms capable of finding solutions to invasive food waste problems and alternatives to unsustainable synthetic materials. The possibility of generating fully biodegradable, compostable and even edible biomaterials is promoting new proactive and sustainable approaches based on the adoption of circular processes.

The use of food in the creation of new biomaterials allows new solutions to be experimented with, often combining traditional techniques and innovative processes in the generation of fabrics or textiles, skins or sheets, surfaces and volumes—more or less flexible, more or less rigid—that can be converted into new bio-products derived from the food industry itself. The resulting products and items—packages, bags or back-ups—can also be edible and tasty items used as natural and safe wrappers.

Experimentation with old and new technologies is making it possible to familiarise and widen the current field of application of biomaterials (bio-plastics, bio-packaging, bio-paper, bio-dyes, bio-ceramics, bio-fabrics) thanks to open-source network dissemination systems, which are increasingly aimed at sharing formulas

M. Gausa and G. Tucci, *Knitting Food: Food and Eco-textiles*, SDGs and Textiles, https://doi.org/10.1007/978-981-97-7582-8_3

and "recipes" among the community of users, researchers, creators or designers, increasing experience and knowledge (Fig. 3.1).

New collaborative alliances between interdisciplinary projects, scientific and artistic crossovers and new design challenges would thus configure the constitution of a new type of sensitivity (and specialty) related to this boom in research on biomaterials and on the (re)design and functional use of organic and food waste [1, 2].

The world of bioplastics, eco-textiles, green packaging and innovative design has opened the door to the (re)use of food waste or discards and the exploitation of new products derived from milk, fruit, vegetables, plastic, mushrooms, coffee, shellfish, algae and many others. This is a sign of the times where eggshells, potato or banana skins or coffee grounds can be as relevant as the set of bits associated with digital systems themselves; systems of high definition and complexity, which, more and more, allow to manipulate and reformulate the old residual wastes into new experimental tests.

Within this framework, various advanced research centres—associated with the development of new informational and digital technologies, interpreted as new interactive systems between data, information, conditions and environments—have focused a large part of their interests in addressing the current importance of food waste (and surplus) as raw materials to develop new materials and derived products: products generated, in general, taking advantage of the new digital devices (3D printing, 3D stamping, 3D extruding, etc.). Centres such as MIT Lab in Boston, Terreform, in New York, Bartlet School in London, Jan van Eyck Academy in the Netherlands, IIT or CNR in Italy, CITA in Copenhagen or BAU School and IAAC in Barcelona (among others) are some of the big "laboratories" that are engaging in this line.

We do not want to dwell on this section but as we have pointed out in other texts the importance of these *milieux*, it is to induce new processes—and proceedings, often still experimental—where the combination of handcrafted and techno-digital industrial techniques (from natural dyes, braided and fibrous yarns or three-dimensional topologies combined with 3D Printing and 3D Stamping technologies) can be combined in new design solutions, processual rather than merely formal (Fig. 3.2).

3.2 Creative Food Cycles: Towards Urban Futures and Circular Economy

In this framework of action, in 2018, the project Creative Food Cycles (CFC), is born: a proposal co-funded by the Creative Europe Program of the European Union, with a clearly European perspective: in line with the own dispositions of the CEE, the project aims to develop a creative and innovative approach to the food cycles, particularly addressing circular economies for positive urban changes; working in several actions,

Fig. 3.1 AI-generated biomaterials (bio-plastics, bio-packaging, bio-paper, bio-dyes, bio-ceramics, bio-fabrics)

Fig. 3.2 Discarded fruit and vegetables are turned into 3D-printed snacks at Upprinting Food. Image: Grace O'Brien

from the research on new bio-materials and bio-products to a set-up of publications, workshops, exhibitions, installations, symposiums and a "festival" oriented to favour the communication of the project's results and the global background linked to them.

Activities developed in partnership and collaboration between the Chair for Regional Building and Urban Planning of Leibniz University Hannover (LUH), the Institute of Advanced Architecture of Catalonia (IAAC, and the GIC.Lab and the Urban and Design areas of the DAD-UNIGE (Department of Architecture and Design of the University of Genoa). Paraphrasing the responsible of the Leibniz team, prof. Jorg Schröder, "Creative Food Cycles was born with the objective to enhance imaginative practices between food, waste, architecture, and conviviality, in a transnational innovative perspective, setting food in the center of cultural discoveries able to materialize and—at the same time—to synthesize ongoing deep changes in citizenship and technology, in everyday behaviours and shifting routines, in scarceness and abundance: a forefront topic to explore and shape pathways of cultural creativity in strong links with social practices.

Food, hence, understood in terms of integrated systems regarding complex sets not only of economic activities, but also of cultural actions and human exchanges that sharply affect long-term sustainability and the living conditions of the cities themselves. With the term 'Food Cycles' novel concepts for the interaction of technological, environmental, and societal forces in food systems and food cultures can be grasped, favoring a driver-concept for a positive change in urban qualities, cultural practices, new urban commons, urban education, as well as overall economic developments, ecological targets, and social integrations" [3].

Promoted and performed by creative groups of researchers in design, architecture and new technologies, the project in order to address food for urban futures, the project was looking for provide understanding, models and practical tools for Creative Food Cycles as a culture-based approach to circular economy with multi-scalar levels of interaction between cities and landscapes, productions and consumptions, *materialities* and *spatialities*.

CFC would cover, in fact, multiple layers of environmental and socio-cultural actions, trying to promote strategic integrations—creative, innovative and fresh at the same time—from the urban and territorial scale to the scale of the design product and the communicative event, in which the factor "food" would be a key indicator as a priority matter and inducing agent of new sustainable and innovative processes at the same time. In fact, as we have pointed out, the new urban-territorial processes need the landscape, as a new operational and relational sustainable and environmental agent that needs, in turn, agriculture as an innovative activity able to preserve the landscape itself.

Agriculture (especially, that developed in low or medium intensity situations) requires, in fact, a new multi-level definition capable of going beyond its primary condition (a programmatic and diversified mixed-use associated with its goods and crops, but also with agro-tourism, km zero hospitality and/or gastronomy, energy generation, digital manufacturing, technical research, etc.) to ensure its own resilient livelihood capacity [4]. These multi-programmatic condition needs evidently FOOD understood not only through its basilar alimentary (and eating) function but as a multi-productive matter; a hyper-matter linked with new circular processes.

The CFC project has approached this complex sequence of simultaneous levels, addressing the term FOOD in 360 degrees: *from production to distribution, from distribution to consumption, from consumption to disposition and transformation,* structuring the project in these three main steps, maintaining crossing objectives, contents and methods in the exchanges set up by the three partners involved in this Creative-Food-Cycle research [5] (Fig. 3.3).

This *three-phase-model* was conceived to stimulate a deeper interconnection among cultural creators, cultural professionals, institutional stakeholders and active citizens, through an open and inclusive approach. In the evolvement of the working program, the different investigated phases would be understood not only as tool-concepts or specialised actions fields, but as interactive systems capable of providing new holistic material and sensory experiences for the Food Cycles themselves, with the aim of establishing a cultural creativity focused on design, and co-design, as main drivers of a deeper understanding, associated with innovative systemic processes called to contribute to a more circular economy.

- The IAAC, Institute for Advanced Architecture of Catalonia, in Barcelona, was more concentrated in the experimentation with the food production phase through the use of new technologies and new generative fields, experimenting also new indoor bioprocesses, for domestic and industrial production.
- The Leibniz University of Hannover worked on the second phase—from distribution to consumption—by imagining temporary and/or pop-up markets that could foster new exchanges between small local producers to easily market and disseminate their products, but at the same time create new spaces and platforms for social relations.

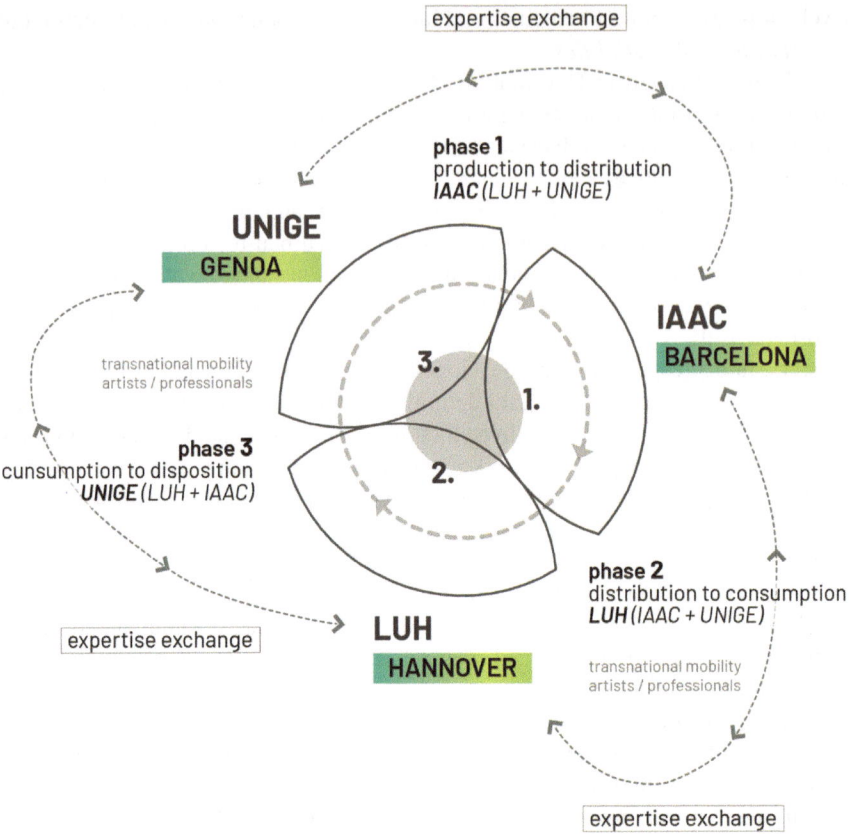

Fig. 3.3 Creative Food Cycles phases and partners' exchanges by UNIGE, LUH, IAAC

- The GIC.Lab-UNIGE with the Department Architecture and Design was more involved in the consumption and the disposition and waste recycle (reproduction) phase, in particular in relation to the food (and food-packaging) waste's reuse by the prototyping of new products and materials, obtained from a second-life food capacity.

The Creative Food Cycles research part developed by the UNIGE team focused on the capacity for proximity production, understood as the capacity for self-production and, above all, awareness of the potential and richness of food waste. As part of the project, on 11 December 2020, the UNIGE team organised the final event of the research: a festival aimed at showcasing innovative new bio-productions and conceived as a menu of shared experiences, sensations, opportunities and social stimuli designed to promote co-participated technical, sensorial and creative interactions.

Food takes on an important "multi-urban" dimension, a new dimension linked, as we have already pointed out, to territory, ecosystems and agro-systems, to new mixed spaces of economic, cultural and social interactions, to new collective scenarios and, finally, to a new design, multi-programmatic, multi-processual and multi-scalar at the same time [6].

The Creative Food Cycle (CFC) program was born with the goal to approach all these levels of action (and activism) trying to promote more complex, informational and strategic innovative dynamics of integration and interaction, reinforcing the different partnerships, researchers and stakeholders that were involved.

Goals in which the factor FOOD has been interpreted as a multi-productive and performative factor able to induce *Wise Urban (and Rurban) Advanced Design logics and prospections* (beyond the simple Smart topic) alluding to this conjugation of systems and subsystems (safety, water, health, mobility, economy, environment, tourism and evidently food), called to orient and manage, in an integrated way, the sustainable development of our new *multi-, inter* and *trans*-hybrid scenarios [7] (Fig. 3.4).

The "CFC International Festinar" (Forum and Festival)—as a new scientific and, before all, sociocultural multi-format—had to be the playful, experiential and experimental culmination of this long *"crEATive"* trajectory, an event in which design should be celebrated and research enjoyed together, near and close; a multiple experience with conceptual installations and selected prototypes based on food waste bio-materials, innovating design models, participated workshops (enriched with the varied contributions of keynote speeches) artistic performances, stakeholders' interviews, exchanges with visitors and social network feedbacks, in an intense open-source interaction, mixing real and virtual formats conjugated in a communicative but, also, expressive and narrative big event [8].

Although it may seem to some that the very concept of a Festival seems far removed from the precision, seriousness and critical severity of scientific meetings; however, a Festival is, in fact, a mixture of exchange, communication and celebration of experiences and ideas. A "festive moment" of collective interactions, expressions, expansions and demonstrations or manifestations (understanding these last, also, as manifestos, that is to say, sometimes, as possible proclamations of ideas and convictions). At every historical moment of change in logics and thoughts—scientific, technological, cultural, economic and socio-political—the celebration of ideas has been as important as their own theorisation, materialisation or dissemination [9].

In the contemporary transition from the post-modern late-industrial age to a new interactive digital time, of *eco-, info-* and *xeno*-logics, more and more networked, material and immaterial, real and virtual, sensorial and *sensorized* [7].

A new dimension in which combination, hybridisation, contamination or impurity is no longer seen as a defect, a deficiency or an imperfection, but as a strategic interactive capacity... and potential.

The global interaction between community and individual seems more sensitive and open to a new virtuous cycle of resources and synergies.

A celebration of the complexity that, once again, translates us into new repertories called to propose more open and carefree formats of narrations, expressions and unexpected configurations. Recovering and reinterpreting, also, the popular models of food delight and interaction (banquets, feasts, agapes, etc.) present in variated

Fig. 3.4 Creative food-cycles event: international festinar "FOOD interACTION" organised by UNIGE team. Image by UNIGE

open formats so characteristics of the Mediterranean atmospheres where sharing experiences and spaces would be so close of the own notion of Festival.

3.3 GIC.Lab Research: Prototyping to Rethink Food Waste

The researches carried out at DAD-UNIGE on biomaterials and bio-products, associated with food and its multiple agro-cultural implications (GIC.Lab and Urban Design laboratories, in collaboration with the DAD-Design area), have been focusing in the problem of food waste—approached from different scales and approaches—and in the awareness of the potential implicit in the new recycling and regenerating processes currently underway, understood as emerging dynamics associated not only with increasingly widespread sustainable production and business models, but also with the possibility of generating domestic examples of self-productive reuse (and, therefore, implicitly self-sufficient reutilisation) with a remarkable degree of experimentation [10].

The current situation of disruptive consumption and invasive waste invites to reflect, more and more, on our own social behaviours and the capacity to process and reprocess food—at all levels and in all its cycles—in the current "urban-territorial" environments where in-between agricultural landscapes are essential to preserve open spaces, create better habitats and facilitate a new type of circular production in the new urban and *rurban* contexts [10–16].

In fact, access to food in the new *polypolis* is not always balanced.

Alongside areas or clusters with abundant (hyper)consumption (rich in plural offers) that often pay little attention to a rational ratio between nutritional need (or delight) and compulsive appetite (or desire), there are entire neighbourhoods that could be defined as "food deserts" [12], in which it is practically impossible to find diversified qualitative products or varied offers of fresh food. Areas which tend to favour, even more, the omnipresence of "fast foods" and repetitive "plastic meals" (so defined by their quality and packaging) with the consequences derived in the accumulation of waste package.

Within this context, the main objective of the experiments and activities promoted by the GIC.Lab-UNIGE team and the training and prospective urban design laboratory has been to design and create concrete and real products and prototypes, starting—precisely—from the valorisation and recycling of food waste in order not only to transmit negative or coercive messages but also to raise proactive proposals in the fight against food waste.

Designers, researchers, planners, technical professionals and entrepreneurs involved in this research have wanted to express, through the development of various materials and functional objects, the importance that a new circular economy linked to the food system can represent, proposing new redefinition systems, in which food waste is relaunched into new productive circuits in the form of new materials, new products or new (re)food. The designs and prototypes generated over these years have been successfully exhibited in various forums and scientific-cultural meetings, often involving the public and visiting experts in an active role through direct experiences, sharing workshops and even creative shows.

The results obtained have also served to reflect on the production, consumption and recycling models present in our habitats, opening the doors to a subject that represents an important topic of current research in the field of innovative design by proposing new solutions to environmental and social problems. Solutions in which systems (and general offers) and behaviours (and particular choices) appear intimately intertwined as demonstrated in the famous exhibition "Broken Nature" at the Milan Triennale [13].

Following the publication of the *Cities and Circular Economy for Food Report* at the World Economic Forum in Davos (January 2019), the Ellen MacArthur Foundation launched the Food initiative [14] driving new food and urban linked strategies:

> Cities - and associated agricultural landscapes - play a crucial role in giving new added value to food (diversifying its essential, primary condition and multiplying its potential destinations) largely eliminating, in turn, food waste [15].

Such scenarios can become essential environments for the redistribution of surplus food (*Second Life Cycles*) and fostering a thriving bio-economy where food by-products are transformed into organic fertilisers, degradable biomaterials, medicines, cosmetic products or bio-energy (Fig. 3.5).

Thanks to the Creative Food Cycles' (CFCs) research project funded by the European Community through its Creative Cities 2018 program, the GIC.Lab/Urban Design (UNIGE) research line developed in Genoa had the opportunity to address this broad topic in greater depth and develop numerous proposals associated with food waste [5, 7, 10, 16].

We have already pointed how the main objective, in addition to the experimentation itself, was to help raise consumer awareness of the potential of recycling by sensitising them, at the same time, to the growing socio-environmental importance of waste and its possible forms of treatment through easily reproducible procedures in everyday use.

Within the CFC project, several food waste reuse projects have been developed with the students, designers, researchers and creatives (Figs. 3.6 and 3.7).

These projects always worked with the main idea of making the population aware of recycling, looking for easy to reproduce procedures and daily use of the products made. There have been different ways of processing waste, but we can say that the three most recurrent macro-categories are the addition of bio-resins or homemade processes that we could define as "cooking chemistry" or situations of drying and weaving of food waste. An excerpt of the processes and products produced can be found in Table 3.1 and Fig. 2.8.

The quality of the products produced by the various research teams involved has generated new methods and open trials, expanding knowledge in the field of design and innovative scientific-cultural dissemination.

FIGURE 4: THREE AMBITIONS FOR CITIES TO BUILD A CIRCULAR ECONOMY FOR FOOD.
In the circular economy vision for food, cities send clear demand signals to support regenerative production and better food design, while turning by-products from food eaten in cities into organic fertilisers for peri-urban farmers to use.
SOURCE FOOD GROWN REGENERATIVELY, AND LOCALLY WHERE APPROPRIATE
DESIGN AND MARKET HEALTHIER FOOD PRODUCTS
FOOD INTO CITIES
INFLUENCE
MAKE THE MOST OF FOOD
BY-PRODUCTS INTO BIOECONOMY

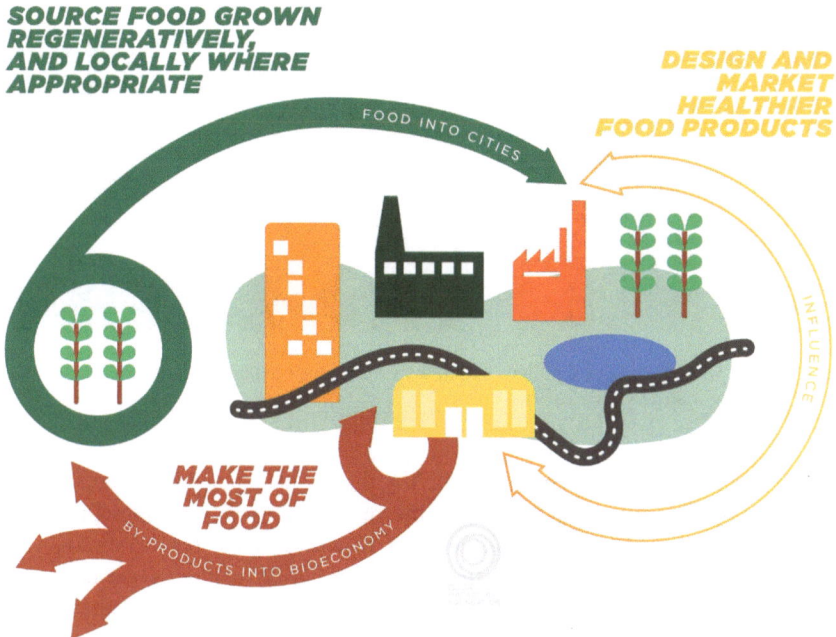

Fig. 3.5 Cities and circular economy for food report, Ellen MacArthur Foundation, 2019

It is possible to identify some macro-categories within which the various food wastes associated with these dynamics have been approached. The main *tecno-artisanal* or *para-artisanal* processes employed in these food experimentations could be summarised as follows:

- "*Drying*" *processes*: the drying processes allow to eliminate (through progressive phases) parts of the initial water contents present in the products, thanks to the administration of heat (ovens, dryers, free air itself). This method is usually used in the initial phase of processing wet waste, such as fruits and vegetables, etc.
- "*Compression*" *processes*: through pressing and compression processes, the treated food acquires a pseudo-farinaceous texture that can be mixed with natural thickeners—such as potato starch, corn starch, gelatines, agar–agar substances—and pressed into rigid moulds in order to define a precise shape. This process tends to be used mainly from (already) dried foods such as flour, coffee residues, eggshells and nutshells mixed with thickeners.

Fig. 3.6 My Upcycling Closet project within the CFC project experimentation by C. Fiorani, C. Ricaldone (GIC.Lab-UNIGE, 2019)

- *"Agglomeration" processes* based on bio-plastic substances: in circumstances of particular need for product strength, bio-resins are used to agglomerate and fix the starting material(s) in a resistant way. This method has been commonly used to create the resistant and joining parts of products, with a great capacity to incorporate medium-sized dry food waste (pieces of pasta or hard pulses, nutshells, fish bones, etc.).
- *"Cooking" processes* (with chemical modifications): concern alterations and transformations of the food chemical components to obtain new material substances: extraction of fats, combination of acid solutions, dissolution of sugars, freezing or cooking, etc., would be some examples.
- *"Manipulation" processes*: refer to products handling or manufacturing transformation in which the initial product changes its form and function, but not its internal composition. For example, the manipulation—in the case of food packaging—of cardboard containers converted into creative elements or the reuse of bottle caps transformed into clothing accessories, as well as pieces of cans used as complements or assemblies in clothes or dresses, etc. Evidently, these last processes aim, to a great extent, at the recycling of packaging and container elements conveniently reformatted.

All these processes are, of course, mixable and combinable, allowing multiple conjugate/conjugate declensions and derivations to be generated.

The results obtained, in line with many similar ones in other experimental centres and institutions, show how it is increasingly possible to reconcile environmental ethical implications, socio-cultural concerns, creative innovation and technological applications, without neglecting the absolutely essential economic development. The

Fig. 3.7 Some of the products and materials designed by UNIGE students on the CFC project, that reflect the list in Table 3.1. Image by GIC.Lab

Table 3.1 Some of the products, materials and processes approached and designed by UNIGE students within the CFC project

Food waste	Type of process	Product description
Coffee grounds	Bio-ceramic compression and agglomeration	**1. Mooka** is a product born from a circular design idea; it is a vase that completes its life cycle becoming fertiliser for the plant. Presented in a setting that offers visitors a visual and olfactory experience
Coffee grounds	Bio-ceramic compression and agglomeration	**2. DishBratta** line is made by mixing coffee ground and a biological resin. It consists of a set of two dishes, a dinner plate and a deep dish, a fork, a spoon and chopsticks
Chamomille infusion	Bioplastic agglomeration processes	**3. BioPlastic** was born from the desire to create a line of packaging for chamomiles and infusions starting from the classic internal waste of the sachets once used
Fennel and walnut waste	Bioplastic compression and agglomeration	**4. Fennut light** is a lamp in two versions, Tube and Pyramid, where two materials born from food waste are used together
Eggshell, pasta, legumes	Bioplastic compression and agglomeration	**5. Bis Bioresina** is based on the experimentation of dry food waste such as pasta, eggs, legumes and thickeners to make reusable dishes and tableware
Rice husk	Bioplastic compression and agglomeration	**6. V.pot** is a dish made from the waste of rice husk compressed in a mould with the addition of bioresins
Fish bones	Bioplastic and eco-textile compression and agglomeration	**7. Bofish** is an innovative material obtained from bone and cartilaginous waste from fish, especially those from tuna caught in the waters of Camogli from tonnarella
Peanut shell	Bio-organic product cooking chemistry	**8. Hanging Plates** starts with the transformation of the peanut shell through a domestic recipe and subsequent baking in the oven into a mouldable material and transformed in this case into bowls
Honey	Bio-organic product cooking chemistry	**9. Miellow** is a honey-based bioplastic. Its strength is resistance and elasticity. It can be used in various ways and is not intimidated by water. Its texture is delicate and fine, the semi-transparency given by honey gives it a glass-like appearance
Milk	Bio-plastic cooking chemistry	**10. Galalith** is a synthetic plastic material manufactured by the interaction of casein and formaldehyde. It is odourless, insoluble in water, biodegradable, non-allergenic, antistatic and virtually non-flammable

(continued)

Table 3.1 (continued)

Food waste	Type of process	Product description
Soybean	Bio-textile fabric drying and weaving	**11. S.D.S.** The skin made of soybean, combined with the weaving process, makes healthy and environmentally friendly coasters and placemats
Loofah	Bio-organic product drying and weaving	**12. Loofah fibre**'s mission is to completely reuse matured and inedible loofah and combine the good physical properties of the loofah

pursuit of real actions and operational objectives is possible through sustainable behaviour towards a circularity of reuse (or recycling) of food-related resources.

In this sense, it is possible to identify some macro-categories within which the various food wastes associated with these dynamics have been approached. Some of the most interesting projects of recent years at international level are the result of experiments on the *second life* of food products which can be synthesised in five categories:

- Biomaterials understood as new matter elements like bio-plastics, bio-ceramics, bio-cements, bio-agglomerates, bio-papers.
- Bio-textiles and derivatives as bio-tissues, bio-fabrics, bio-leathers, bio-skins, bio-films, bio-dyes and bio-pigments.
- Bio-products as industrial or semi-industrial design issues like home accessories and furniture (tablewares, plant-pots, lamps and light elements, chairs, tables, panels), jewellery, craft papers.
- Bio-structures as new *naturartificial* hybrid structures composed of one or more biomaterials.
- Bio-packaging and associated packing or containing items.

The sectors of textile, lighting, ceramics or construction industry, plastics and paper, waste tableware or packaging, as well as the wider fields of design and architecture, furnitures and jewellery, have opened their doors to new products derived from milk, fruit, vegetables, plastics, mushrooms, coffee, seafood, algae and many others.

Within this context, the broad field of eco-textiles based on food waste is particularly relevant: eco-textiles, bio-fabrics, bio-skins and bio-wovens today represent an important category of biomaterials that have immediate applications in sustainable industrial production and the circular economy (Fig. 3.8).

Fig. 3.8 GIC.Lab. Fruit Leather project. Designers: Federica Vicini, Luigi Scala, Noemi Campion

References

1. Magee C (2020) The use of food waste in the development of biomaterials. https://www.inn
   ovationnewsnetwork.com/the-use-of-food-waste-in-the-development-of-biomaterials/6697/
2. Markoupoulou A et al (2019) Food interactions catalogue. Collection of best practices. IAAC,
   Barcelona.
3. Schröder J (2019) Creative food cycles towards urban futures and circular economy. In: Markoupoulou A et al (2019) Food interactions catalogue. Collection of best practices. IAAC,
   Barcelona
4. Pericu S, Gausa M, Tucci G, Ronco Milanaccio A (eds) (2021) Creative food cycles experience.
   Goa CFC-festinar: a virtual banquet for an innovating research celebration. ADDDoc Logos,
   Genova, pp 13–52
5. Gausa M, Pericu S, Canessa N, Tucci G (2020) Creative food cycles: a cultural approach to
   the food life-cycles in cities. Sustainability 12:6487. https://doi.org/10.3390/su12166487
6. Tucci G (2020) MedCoast AgroCities. New operational strategies for the development of the
   Mediterranean agro-urban areas. Trento-Barcelona, ListLab
7. Gausa M (2021) CFC–multiscalar challenges. In Pericu S, Gausa M, Tucci G, Ronco Milanaccio
   A (eds) Creative food cycles experience. Goa CFC-festinar: a virtual banquet for an innovating
   research celebration. ADDDoc Logos, Genova, pp 13–52
8. Ricci M (2012) Nuovi Paradigmi. Trento- Roma-Barcelona, List Lab
9. Gausa M, Fagnoni R, Galli G, Bilancioni G, Falcidieno ML, Prati F, Vannicola C (eds) (2013)
   Rebel matters-radical patterns. II International Congress, ed. GUP, De Ferrari Editore, Genova
10. Ronco Milanaccio A, Tucci G (2021) GOA Reasoned recipes book. Prototyping, experimenta-
    tion and innovation to rethink the food waste. In: Pericu et al (eds) Creative food cycles expe-
    rience. Goa CFC-festinar: a virtual banquet for an innovating research celebration. ADDDoc
    Logos, Genova
11. Cockrall-King J (2012) Food and city: urban agriculture and new food revolution. Prometheus
    Book, New York

12. Dutko P, Ver Ploeg M, Farrigan T (2012) Characteristics and influential factors of food deserts, ERR-140, U.S. Department of Agriculture, Economic Research Service, August 2012. https://www.ers.usda.gov/webdocs/publications/45014/30940_err140.pdf
13. Antonelli P, Tannir A (2019) Triennale di Milano. In: Broken nature. La Triennale di Milano Electa, Milano
14. Ellen MacArthur Foundation (2019) Cities and circular economy for food. https://www.ellenmacarthurfoundation.org/our-work/activities/food
15. Calori A, Magarini A (2015) Food and the cities. Edizioni Ambiente, Milano
16. Gausa M, Canessa N with Tucci G (ed) (2018) Agro-cultures, agro-cities, eco-productive landscapes. Actar Publishers, Barcelona-New York

Chapter 4
Eco-textile as (Cr)edible Matters

4.1 Biomaterials and Eco-textiles: *Eco-* and *Pro-*active Processes

Since ancestral times, man has accumulated and superimposed layers of skins and fabrics on his skin to better manage his relationship with the environment [1].

Wearing clothes—getting dressed—is an exclusively human characteristic. As early hominids migrated towards colder climates, they began to turn to animal skins to protect themselves from the weather [2].

However, since the Paleolithic, about 60,000 years ago, humans had begun to braid threads of plant materials to make strings and ropes. They were also used in animal skins, which were sewn with said primitive cordage or with tendons and bone needles from the same prey. Around 8000 BC, certain pre-Columbian peoples began to develop cotton spinning techniques and around 5000 BC, the treatment of cotton starts in India, the Middle East and Egypt with the creation of their own tools for spinning and braiding, such as combs, spindles (used to convert raw materials into thread) and looms (used to weave cloth and cloth) [1].

The perfection of the primitive vertical loom helped produce the first fabrics. They had a very basic structure—aided by counterweight pieces made of stones or ceramic elements—and allowed the creation of the first textile clothing obtained by elemental mechanic devices. There were also other manual making processes such as basket weaving, which helped in the processes of collecting raw materials.

In the Neolithic era—between 6000 and 3000 BC—the sedentary nature of human beings favours agriculture, livestock and crafts. The domestication of animals made it possible to obtain wool, one of the most common natural fibres. Agriculture would make it possible to grow, in addition to fruits and grains, other plants with a more exclusively textile purpose, such as flax or hemp.

Egypt has often been considered the precursor of the first spun yarn fibres brought about by the discovery and processing of flax, already used by the Mesopotamian peoples as the main spun material. It was, in fact, in Egypt where the first iron needles were created in the year 2000 BC, with a very closed hook through which the thread was introduced [1].

The distaff or spinning wheel appears in China in 3000 BC, favouring greater precision in the compression and reduction of organic fibres converted into threads mechanically, with devices based on wheels, cranks, pedals and supports or rotating axes to accumulate the thread. The spinning wheel made the process of spinning fibres easier, improving their characteristics and facilitating their adaptation to transformation into fabric. For both domestic and commercial use, the creation of fabrics, through different textile fibres was based on these two machines, which underwent different evolutions to optimise manufacturing.

The improvement of the spinning wheel was going to favour, in turn, the treatment of silk, a precious textile material of animal/plant origin (silkworms and mulberry trees) that spread to the Persian Empire, finally reaching Europe along the famous Route of Silk. The arrival of the industrial revolution completely changed the world of textiles; the textile industry chose to abandon traditional forms of handicraft production, realising that technological innovations allowed for increased production.

The creation of the first steam-powered spinning machines (powered by man and animal) and multi-arm mechanical lathes were some of the most significant advances between the 18th and 19th centuries [1].

Until the mid-nineteenth century, clothes were made in homes by domestic and artisanal means or to order, by local tailors, and sold through exhibitions. Access to the main tailors' shops, with more sophisticated fabrics, was left to the newly emerging bourgeoisie or aristocratic customers. Soon, the great "trendy" fashion designs, always produced locally and on a relatively small scale, would be paraded on the catwalks of major European cities [3] (Fig. 4.1).

After the second industrial revolution and the spread of electric machines, textiles made a quantum leap in its production and socio-economic dynamics.

The "culture of consumerism" took hold around the 1950s, thanks to the increase in mechanisation and the growing prominence of advertising through television after the Second World War [3].

Soon, natural and man-made fibres were mixed with synthetic fibres and chemical industrial processes developed polymeric materials such as nylon. Similarly, more complex products, through polycondensation processes with different reactive agents, created ultra-strong polymers with textile applications, resulting in fabrics such as Kevlar, five times stronger than steel and able to withstand the impact of a bullet without being pierced. Following the style of this fabric, more and more textile materials were produced that could withstand various chemical and mechanical processes.

Industrialisation brought textile mechanisation closer to homes with the creation of the sewing machine. This household tool (popularised by the Singer company with its innovative mechanical patent with a rectilinear motion shuttle and an uncurved needle) facilitated domestic sewing tasks in the Western world.

Their contribution made sewing machines an easy-to-use tool, which is why they became a typical element of any home [1].

Fig. 4.1 Famous photo of Gandhi with his spinning wheel as a symbol of self-economy

Using clothes as an extension of our skin is, hence, a daily fact inherent to the evolution of our technological-cultural progress.

Traditional materials such as leather, silk, cotton and wool (all products of nature) became the basis for most of the textiles used in the course of time (Fig. 4.2).

However, during the twentieth century, the industrial revolution and chemical advances helped replace these materials with synthetic substitutes (acrylic, nylon, polyester, etc.), which were easier to produce and designed with greater precision and efficiency. These synthetics favoured cheaper and more functional processes, but generated, in return, increasing environmental pollution and immense waste of materials [2].

Since the 1970s, most clothing production has moved abroad and the scale of production and speed to market has accelerated. Fast fashion took hold in the 1980s and some have called it the "democratisation of fashion". What once seemed exclusive to the few became accessible to the majority [3].

In a global scenario of increasingly ephemeral, fast and disposable fashion, the textile industry has had an increasing impact on the environment due to the use of toxic chemicals, water and energy consumption and heavy logistical transport. The fashion and textile industry is the second most polluting industry on the planet according to the United Nations and is responsible for about 10% of CO_2 emissions and about 20% of wastewater [4].

Every time we wash a garment, up to 700,000 microscopic fibres reach the oceans. Thirty-five % of the microplastics found in the sea come from synthetic fabrics. These figures, added to the fact that the world population is progressively increasing and

Fig. 4.2 Weaving factory Colònia Sedó in Esparraguera, Catalonia/Spain about 1900. *Source* MNACTEC, https://museucoloniasedo.cat/ca/galeria-dimatges/

that 20 new garments per person are produced every year, force us to think about a necessary paradigm shift that starts from the initial approach to production.

More than 80% of the environmental impact of a garment is in fact a consequence of the design process. The challenge will be to rethink all phases of the product's life, to favour a circular economy. Designing with recycling and complete biodegradability of each garment in mind is therefore essential [4] (Fig. 4.3).

The use of biomaterials (and in particular of eco-textiles) from nature (animal or vegetable) or from the recycling of food itself (*second-life productions*) has been strongly emerging in recent years.

It is now possible to obtain textiles from pineapples, milk proteins, algae, oranges, coffee, mushrooms, apples, cacti, spider webs, etc. Recycling and the use of environmentally friendly textiles are currently one of the clothing industry's priorities to ensure a more sustainable development for the future, as consumers today buy 60% more textiles than two decades ago.

Fast-fashion offers an immense amount of "seasonal" clothing at cheap prices. Although the big brands claim they are changing their production systems, the truth is that today there is still an uncontrolled "supply/demand" ratio. The future of fashion undoubtedly foresees an eco-innovative development, but also a change in consumer habits and behaviour, aimed at buying less and more responsibly.

Today, research is increasingly focusing on the development of sustainable, biodegradable, reusable and, to a greater or lesser extent, intelligent (responsive, adaptive, sensitive) materials and eco-fabrics.

Fig. 4.3 Current textile processes with fast, automated machinery. *Source* ED Textile, https://fdt extil.es/procesos-y-maquinaria-en-la-industria-textil/

The old industrial warehouses of mass production give way to (or are supplemented by) new experimental and interdisciplinary laboratories that demonstrate the need for new textile research [3] (Fig. 4.4).

This new techno-sustainable sensibility involves companies, designers and researchers in tackling the growing environmental problem, allying themselves with biology, chemistry and new production technologies, to create projects capable of exploiting natural resources in a circular manner, recycling waste (organic, food and obviously textiles but also packaging) in many ways to achieve greater biodegradation capacity and minimum contamination: but also capable of adapting, reacting, evolving, growing and even repairing [2].

As we have already mentioned, many emerging biotechnology or bio-design companies are developing new biomaterials for the textile industry through cell cultures of biopolymers, macromolecules (such as cellulose) or through bacteria and proteins present in living organisms, as well as through the use of food remains.

The result is thus new textile biomaterials (in particular fabrics, skins, films, combinable pieces or accessories and complements, etc.) that not only decompose naturally at the end of their life cycle, but also have various additional qualities, such as the ability to absorb carbon dioxide from the air.

Eco-textiles are generally, in fact, structures composed of fibres derived from various sources that are subsequently processed and transformed into knitted fabrics, flat fabrics, nonwovens, etc.

Currently, like other biomaterials, they are also used in the medical field (hernia repairers, arterial grafts, artificial skin, ligaments and tendons, surgical sutures, etc.) [5].

Fig. 4.4 Digital fabric laboratories and Fab.Labs. *Source* IAAC, FabriAcademy, 2021, https://tex
tile-academy.org/

The notion of eco-textiles is not entirely new in this field: the first steps date
back to the 1930s, when "Lanital", a fabric made from casein, a milk protein, was
marketed by SNIA Viscosa in Milan.

Subsequently, as we will see on these pages and in the product catalogue (Chap. 5),
research has broadened the field of experimentation, which has progressively led to
greater sophistication in biomaterials and more specifically in that of eco-textiles.

However, eco-textiles have never achieved widespread popularity to date because
major brands and retailers have preferred the standardised industrial processes
of traditional natural fibres or synthetic materials. However, new social and
environmental awareness is rapidly changing trends [1].

Bio-design is proposed as a current future in the new eco-industrial processes of
textiles and fashion, and many experimental proposals are already ready to be used
to overcome current conventions.

Experimentation and research on eco-textiles—as part of a broader process associated with biomaterials—change our relationship with industry and challenge us to
think about new logics of production and consumption.

Now, the collaboration between designers, scientists and new technologies will
be the key to lead us towards a more advanced and circular industry and economy
(Fig. 4.5).

Fig. 4.5 Algae Experimental Skins, Atelier Luma and Klarenbeek and Dros, France-Germany, 2017. *Source* https://www.dezeen.com/2018/09/27/the-rising-use-of-recycled-plastic-in-design-is-bullshit-says-jan-boelen/

4.2   The Role of Eco-textiles Within the Advanced Design

At the end of the 1990s, more precisely in 1997, the mobile phone company Ericsson launched an advertising campaign in which various models appeared dressed in "strange" clothes made of climbing plants, natural and/or synthetic grasses, fish or an octopus suitably adapted—like a second skin—to the user's body.

The project had a strong media impact but, above all, it was a source of inspiration—widely reproduced in theory and design books—for a new kind of pioneering actors in architecture and design who experimented with intersections between long-separated dualities: nature and artifice, landscape and city, raw material and refined construction, etc. [6]. The possibility of an interactive, celebratory and expressive hybridisation between new combined realities (multiple situations, conditions and information, but also people, habitats and environments) gave rise to new emerging logics of a digital revolution more open to a processual complexity. A multi-layered complexity manifested directly, without formal, aesthetic or linguistic preconceptions, interconnected and open to new research paradigms, a manifesto of creative freedom summarised in the slogan *"express yourself"*.

Curiously, the series of four advertisements was based on a fusion between the textile universe and a raw nature (manipulated and respected at the same time—synthesised in vegetable or animal coverings) associated with a technological element (here symbolised by mobile devices) acting as a catalyst and even a virtual mediator.

After almost 40 years, the first pioneering and anticipatory experiments in the hybridisation of body, garment, fabric and natural organisms have been largely consolidated thanks to the advances of new information technologies in the world of adaptive design, 3D printing, reactive micro-organism sensors, advanced materials chemistry and digitised process management (as well as a new sensitivity to the environment and its responsible approach).

As mentioned above, within this innovative and experimental context, the field of bio-textiles derived from food waste is particularly relevant as it now represents an important challenge towards sustainable industrial production and the circular economy.

We have already pointed out how the equation "BIO-technology + AGRO-culture + SOS-tenability + HYPER-materials" (or Materials + Science + Digital Fabrication + Environment) is increasingly decisive in this field [7].

Recently, this awareness has generated—in the world of research and industry—new concepts and definitions such as "advanced design", "material ecology", "generative design", "organic design", etc., which combine skills and tools at a multidisciplinary level in the fields of materials engineering, architecture, design, information technology, chemistry and biology.

Around this multidisciplinary approach, the new eco-techno design is being developed with the aim of combining the world of textiles with that of research, as both creative and scientific experiments have developed over the last decade.

The rise of fast fashion and overproduction in recent years has had devastating social and environmental consequences. It is crucial that the textile and fashion industry explore more sustainable and viable production and distribution practices in order to preserve a healthy ecosystem for future generations [8] (Fig. 4.6).

The creation of Fab.Labs, for instance, (laboratories for experimental prototyping and advanced production design) represents a first innovation process in the field of digital fabrication. Related to the Fab.Labs network, the *Fabricademy* (founded by Anastasia Pistofidou) represents an international network of innovative textile research, focused on the application of emerging technologies to optimise the fashion and textile industries (in the areas of creation, production and distribution issues), reinventing alternatives to industrial processes through localised supply chains, open-source methodologies and knowledge exchange, also working in the areas of wearable devices and technological response [8].

The concept of "biomanufacture" proposes that living organisms (bacteria, fungi and yeasts, among others) function as fibre factories. Fermented and cultivated in the laboratory, they can be genetically modified. A pioneer in this type of study is North American fashion designer Suzanne Lee, creative director of Modern Meadow, a start-up that develops Zoa, an imitation skin based on the modification of yeast cells that generate a specific collagen with a specifically designed DNA and allow fermentation and germination of "(re)producing" cells that, through various purification and natural dyeing processes, results in a texture similar—though lighter—to skin.

It is true that many of the bio- and/or eco-textiles currently under development are still prototypes. In most cases, they cannot yet be considered consumer products because there are no guarantee protocols yet in place for strength, flexibility and durability, among other qualities [9].

Verónica Bergottini also moves along the same line of developing bio-fabricated prototypes with her *BioTex* proposal, a product based on materials of microbial origin to create wallets, envelopes and other accessories, as well as textiles with applications of bacterial nanocellulose obtained through the propagation of yeasts and bacteria called Scoby (an acronym for Symbiotic Colony of Bacteria and Yeast), as well as apple cider vinegar, sugar and tea bags. The production, carried out in a techno-artisanal manner, faces a great challenge in experimenting with sugar-rich industrial waste (malt, wine, yerba mate, etc.) (Figs. 4.7 and 4.8).

Another significant achievement is the material created by the multidisciplinary team (engineers, designers, architects, biologists, researchers) of the Tangible Media Group of the Massachusetts Institute of Technology Media Lab (MIT), together with the Department of Chemical Engineering, the Royal College of Art and the sportswear company New Balance. *BioLogic* is a fabric composed of bacteria that interact with the movement generated by the body of the wearer wearing the material.

The development begins with microbiotic cells cultured and harvested to generate a concentrate that is microprinted onto the latex, resulting in a biohybrid film that reacts to sweat as the user (athlete or dancer) performs their activity. The new material is equipped with small flaps that open when one sweats and close again when the body stops sweating.

The prototypes developed so far have focused on running clothes and shoes. The goal is to keep the user dry and cool, and the most ambitious next step is that this innovation also neutralises odours caused by sweating or perspiration [9].

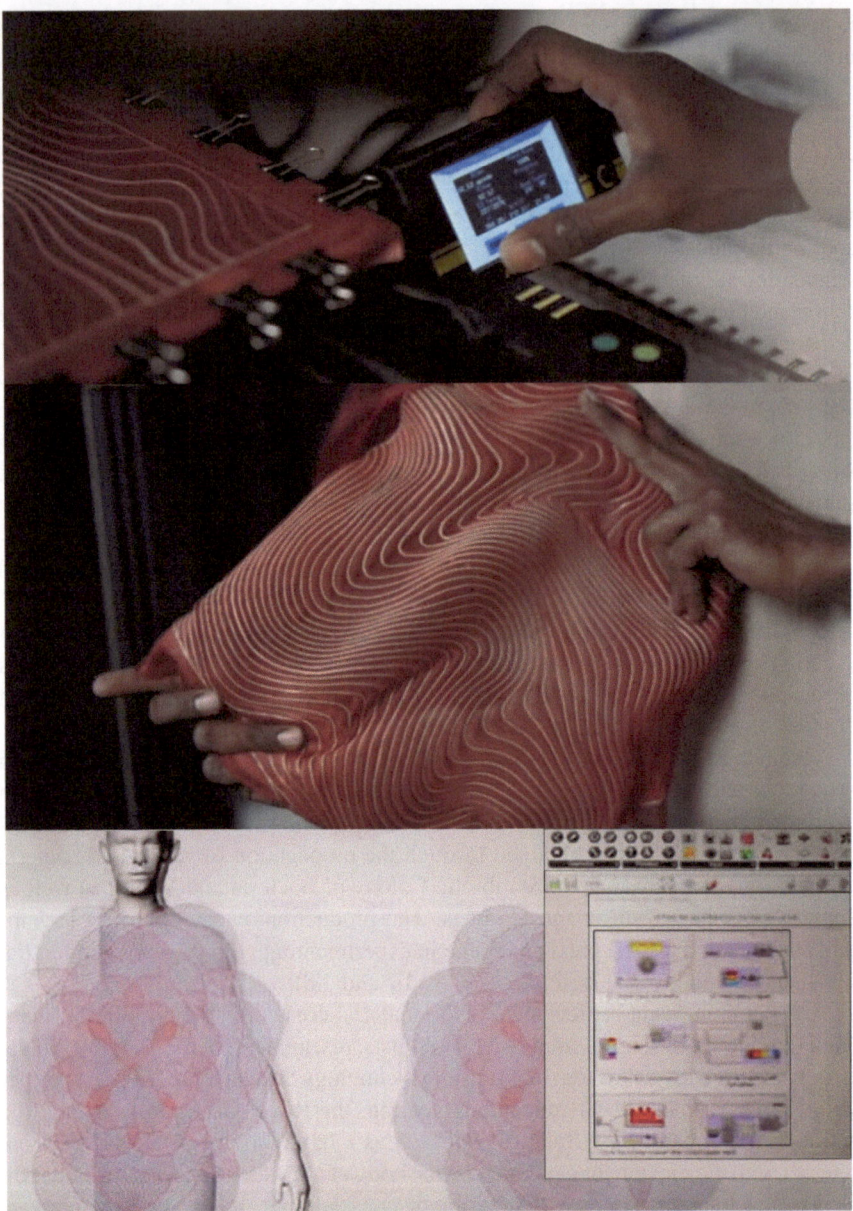

Fig. 4.6 IAAC-FAB LAB-FabriAcademy: *Expanded Lines*, by Dinesh Kumar (2022). A responsive textile material made by digital 3D fabrication. https://www.youtube.com/results?search_query=the+possibilities+of+3D+printed+fabric

Fig. 4.7 Veronica Bergottini: bacteria cropping for yerba mate nanocellulose tissues. *Source* https://agendarweb.com.ar/2019/09/29/una-cientifica-argentina-creo-un-material-con-yerba-mate-para-aplicar-en-la-moda/

Although there are still many doubts and questions to be resolved, the main contribution of this research is the generation of materials aimed at significantly reducing pollution, favouring a lower impact on the environment (reduction of water use and CO_2 emissions) and a new "proactive contract" with Nature.

One of the most representative figures of this new approach and thought and of this "new revolution" is the artist and architect Neri Oxman, founder of the new discipline called "material ecology", whose idea is based on the relationship between design and nature, where nature has the role of helping in the creation of materials that are self-producing, grow and are also able to repair themselves, like a human body.

Hers is a hybrid, multidisciplinary approach in which design and architecture do not impose themselves on the environment but are part of it.

Today, perhaps under the imperatives of growing recognition of the ecological failures of modern design and, inspired by the increasing presence of advanced fabrication methods, design culture is witnessing a new materiality.

In the last decade, in both industrial design and architecture, a new body of knowledge is emerging within architectural praxis where the fact to Design (with) Nature implies moving towards a new kind of processes (mixed with new kind of matter-fabrics). With the growing relevance of a new "bio-materialization", the porous boundaries between material science and digital fabrication are supporting the

emergence of new perspectives in industrial design based on the crossover between various disciplines disciplines (from industrial design to textile design) making computational design at the forefront of research.

We are on the cusp of a new paradigm inspired by the Troika structure of craft, at the interaction between Materials, Science, Digital Fabrication and Environment. This new Bio-Material Ecology defines an emerging field in design, denoting a new logic in the informational relationships between products, habitats, systems, and environmental conditions (Neri Oxman) [7].

In the field of neo-clothing and neo-textiles, Oxman and the Mediated Matter Group at the MIT Media Lab have gone as far as creating new textile materials that

Fig. 4.8 Tangible Media Group & MIT. *BioLogic* a biohybrid film that modifies itself, reacting to sweat and at temperature. Image. https://arts.mit.edu/biologics-living-textile/

can interact with the environment and support humans in life-hostile environments (Fig. 4.9).

Using bioresins and bacteria, they envisioned the Wanderers Collection (2014), a series of innovative 3D-printed garments designed for interplanetary travel and engineered to host synthetically engineered microorganisms capable of generating sufficient amounts of biomass, water, air and light to sustain life in harsh environments.

The project was defined, implemented and explored by the author in the context of new design works commissioned by the Centre Pompidou (Paris).

Within the collection, 18 prototypes were conceived and fabricated for the human body, such as human garments-armour and "textile neo-structures", suitable for multiple functions, including new types of vests, corsets, helmets, armbands and various hybrid devices, which improve resistance, promote innovative and flexible comfort, or explore functional combinations between the growth of embedded performance organisms (algae, bacteria, sedimented liquids of colour, etc.) and bioplastic evolutionary fabrics—skins or leather [7].

Through photosynthesis, the microorganisms, which inhabit the 3D structures, convert light into energy, others with bio-mineralisation strengthen and grow human bones, and still others with fluorescence light the way in the dark.

Each garment transforms elements found in the atmosphere into one of the classic life-supporting elements: oxygen for breathing, photons for seeing, biomass for eating, biofuels for moving and calcium for building bones. The design research at the heart of this collection lies at the intersection of multi-material 3D printing and synthetic biology.

The trinomial "eco-textile/sustainability/technology" but also "materiality/digital manufacturing/environment" are thus fully merged to give life to a new innovative way of rethinking the world of architecture, design and fashion, which—as will be seen in the bio-textiles catalogue 4.2 collection—has found and continues to find very fertile ground in the creativity of professional-designers and in the increasingly conscious choice of pioneering R&D companies that decide to favour new bio-materials and bio-fabrics for their productions.

With the Wanderers collection, this approach seems to end as it began with the Ericsson advertisements of the 1990s, with a set of strange hybrid garments expressing the need for a new creative freedom (more responsible) and a new contact with nature (more reactive).

Fashion accessories that, despite seeming out of place, actually express the new eco-technological approaches, currently underway, in the field of design in general and textiles in particular.

As we will see in the "Knitting catalogue" of this work (Chap. 5), this new paradigm of innovative and sustainable production in the fashion industry is taking shape not only at the level of experimentation and prototyping but also at the level of professional possibilities.

Proof of this global change are the numerous companies and start-ups that have now created a business in the eco-textiles sector, i.e. *Mango Materials*, which produces biodegradable natural bio-polyester fibres as opposed to petroleum-based

Fig. 4.9 Wanderers, an Astrobiological Exploration, Neri Oxman + Christoph Bader and Dominik Kolb, 2014. *Images* https://neri.media.mit.edu/projects.html

fibres; *Ecovative Design*, which works with fungal mycelium to generate a fabric with skin-like properties; *Bolt Threads*, which generates a micro-silk, based on living organisms, emulating the fibres created by spiders (Stella McCartney used this material to develop a garment presented at the Items: Is Fashion Modern? exhibition at MoMA).

These proposals are joined by *Wissahickon Shoe*, a prototype designed by Silvio Tinello (which combines two material techniques, a fungal bio-agglomerate produced from mycelium and yerba mate waste for the sole and a bio-leather for the upper and moulding); and, again, *Piñatex*, which uses pineapple waste from leaf fibre to obtain a living bio-textile subjected to an industrial process already adopted by the Puma and Camper brands [9].

Europe is also recognising the importance of promoting these innovation processes and training new generations in this area. Recently, for instance, an Erasmus + project—re-fashion.eu—has been financed dedicated to the creation of a new vocational training course and a corresponding vocational skills/competence profile, in line with ECVET, capable of assessing, identifying and applying sustainable solutions, shaping a comprehensive sustainability strategy focused on more sustainable raw materials, sustainable production practices, eco-design of textile products, improving the durability and life cycle of products, facilitating end-of-life management of textile products and implementing circularity practices in the textile and fashion industry.

The project also aims to help SMEs and consumers cope with the expected regulatory and market challenges for the fashion and textile industry by increasing their capacity and knowledge through awareness raising and education, to create a new positive and conscious impact in the field of fashion and design.

4.3 Food By-Product Textiles and Circular Design

As illustrated in the following catalogue (Chap. 5), the field of biotextiles in the last decade has shown how creativity, combined with experimentation and the development of new skills (engineering and chemical) on the composition of materials, has initiated an innovative virtuous process capable of radically changing the design-sustainability- circularity relationship by rethinking the food system.

In this sense, the contribution of these international researches to evolve new paradigms for a more *Circular Design* [10] is based on the understanding of food as cross-cutting field of innovation for circularity, crucial for transition [11] and to: design out waste and pollution, make products, architecture and cities regenerative by design and enhance natural capital [12], as well as to support a renewal of community aggregation and new economic opportunities, triggering creativity, technologies, knowledge and abilities towards a "performance economy" [13].

Thus, *Circular Design* embodies the need of the city for effectiveness and adaptivity of strategies, tools and processes of change. Linked with the understanding of design as mode of research and of research as part of design, *Circular Design* can lead to promote design itself, not is products, as culture in being for the age of climate change.

Through methodological innovation, it responds to the need to bridge between the traditional fields of product and communication design and architecture and urban design, to connect between scales and to offer a new setup for cooperation with other disciplines [10].

The "spatial-digital nature", at the very core of *Circular Design*, enables this new mission and role of design. A "learning nature" of multiscalar design processes, from the city to products, is shown in the three main ingredients that shape a creative interpretation of urban contexts (spatial, social, environmental, technological and economical) to promote new structures, networks and connection systems.

Co-production and co-creation are fully involved in an extended framework associated to a new digital (global and local) knowledge, in a time of transitions where the understanding of advanced info- and eco- technologies as multidimensional tools (for design, production, interaction, etc.) implicates new spatial-relational qualities (meshed and netted) within cities and *rurban* scenarios [14].

In support of this new line of thinking, eco-textiles implement a radical approach to create new meanings for waste and "circular" processes, where food and food waste become regenerative (and recyclable) natural systems, turning into evolving resources.

"Making the most of food" [12] means involving local communities, stakeholders and the active urban society, developing a cultural and holistic approach for different *(t)issues* linked to the field of biomaterials, in general, and bio-fabrics, in particular.

It also means uniting all aspects of food cycles and stimulating (with an open and inclusive approach) a deeper interconnection of disciplines and processes dealing with a more cared environment and a more sensitive (co)design.

To do this, we need to start again from food waste, putting it at the centre of a new Second Life (re)Cycles as a primary resource.

The aim is to valorise the role of waste in the creation of new products and services that can be of value in activating creative communities and promoting sustainable behaviour [15] capable of overcoming the "take-make-dispose economy" [16] resulting from decades of consumerism.

The concepts of food waste, food loss or food surplus are not only topics of debate, but also powerful tools for raising awareness of sustainable development at the community level.

Climate change requires original and radical thinking, and while design is a vital form of political action, designers play an important role as powerful agents of change [17, 18]. If we are to pursue what Escobar calls "futures that have a future" [19], we need greater civic awareness.

We must use design as an enzyme to transform the way we live, educate, practice. One of the most important aspects of progress is that it is not about the accumulation of objects, technologies and knowledge: it is about awareness [20].

Humanity's survival depends on understanding ecology and live accordingly [21].

- If food is a shared need (of alimenting), it is also a shared way (of thinking).
- If fabric is a shared need (of covering), it is also a shared way (of re-connecting).
- If food by-product materials are shared potentials (of reusing), they are also shared ways (of re-designing).

In line with the Circular Design approach, within the Creative Food Cycles research, described above (3.2–3.3), the GICLab Urban Design team, in collaboration with the young creatives and designers involved in the project, developed experiments and prototypes of new eco-textiles and derivatives, aimed at the recovery and valorisation of food waste and packaging. Below are some of the most significant projects (Table 4.1; Fig. 4.10).

4.4 Urban Outcomes: Eco-textiles in Cities, Spaces and Communities

As Robyn Metcalfe argues in his recent book Food Routes [22], the household economy has been transformed by the constant search for new fresh foods, as well as their convenience and customisation to satisfy consumers' food desires and phobias: peaches from the farmer next door, a hundred types of bread, spices and exotic fruits for all seasons, dozens of types of coffee, long-life milk and large quantities of meat [23].

The figures are staggering and see the city as the pivotal context for the action.

Every day, in urban areas as large as London, Paris or Berlin, more than 30 million ready-to-eat meals are produced, processed, transported and stored, of which almost 47%—a footprint of 0.74 kg/day per person—is wasted, with no recycling strategy in place [24, 25].

In the Hungry Cities, as named by Carolyn Steel [26], people's perceptions are far removed from the real knowledge of food production, distribution and consumption systems, where it is taken for granted that the food supply in a shop or restaurant is continuously replenished day after day.

The scenario of action within which the dynamics related to the Food System develop is therefore interdependent on the urban space, and the dynamics affecting this system have the capacity to transform the city, activating social, economic, cultural and environmental benefits, or conversely impacting on them.

The need to pay more attention to the role of urban space in the 2030 urban agenda is linked to the challenges posed by rapid urbanisation, towards sustainable transport patterns in the era of digital transition and post-metropolis.

In this sense, the urban space returns to the centre of planning, renews itself and, with a new innovative dimension of networks and strategies, becomes the stage for the new "food issues" that are increasingly central to a more conscious cultural and behavioural transition [27].

Design disciplines, in fact, can support the urban community in building the places of its own interaction with food through multiple occasions of social innovation

Table 4.1 Different food by-product projects of the GICLab-Unige in the field of the eco-textiles (eco-leathers, eco-dyes, eco-skins)

1	**My Upcycling Closet** *Designers: Chiara Fiorani, Carlotta Ricaldone*	Food packaging is transformed into futuristic wearable accessories to enhance body parts and at the same time give value to packaging by turning it into works of art. Bottle caps and tetra pak wrappings are transformed into textiles to create fashion accessories that make people think and draw attention to contemporary consumerism and the environmental issues of overproduction
2	**Prickly Cycles** *Designers: Lorena Likaj, Federica Pelle, Kuo Yu*	The project consists of recycling every part of the prickly pear as a primary material. By experimenting with the fruit waste, it was possible to create a new type of organic material that is 100% compostable, with a varied texture and malleable consistency. This new organic fabric material was then applied and integrated into fashion accessories such as bags and necklaces
3	**Ri-carta** *Designers: Erijon Ademi, Lorenzo Decia, Chiara De Filippo*	Ri-carta is a project inspired by paper and cardboard recycling techniques. The aim is to recycle paper food packaging to obtain sheets of various sizes with interesting and different textures. Once assembled, it is possible to obtain wearable garments, such as waistcoats, hats, papillons, or, by combining them, modular installations. Thanks to the particular translucence of the material, beautiful light effects can be obtained
4	**Drink Dress** *Designers: Enrico Levo, Margherita Massalin, Giacomo Rossi*	The DrinkDress project consists in the development of a dress entirely made with aluminium cans and recycled metal rings that act as elements of union between the elements that make up the dress. This project wants to sensitise people to the reuse of materials that we consider single use. The dress is made up entirely of scales of different sizes that allow an excellent adherence to the body and the dress can be easily adapted to different sizes thanks to the back zip that allows you to tighten or widen the garment as needed
5	**Giano** *Designers: Federica Vicini, Stella Femke Rigo*	Giano project has its roots in the food waste of fruit and vegetables, focusing on the huge amount of produce that is thrown away every day because it is overripe or damaged during transport. The aim is to find a solution to reprocess the waste elements and channel them into the production of a series of natural dyes for colouring textiles, paper and wood. Some of the waste materials used in the creation of the natural pigments were: raspberries, carrots, avocado, cherry, onions, apples, beetroot, pomegranate, blueberries
6	**Fruit leather** *Designers: Federica Vicini, Luigi Scala, Noemi Campion*	The project is based on finding a solution to avoid mixed fruits and vegetables waste and rework it into an innovative product capable of giving these fruits a second life. By drying, re-pulping and processing excess, damaged or spoilt fruit waste, it is possible to produce sheets of bio-fabrics with a soft and pliable texture. The sheets produced have excellent textural qualities and, combined with imagination, can be used in a variety of areas, from setting and decorating a table to creating accessories and clothes for wearing

Images https://gup.unige.it/creative-food-cycles-experience

Fig. 4.10 Different food by-product projects of the GICLab-Unige in the field of the eco-textiles (eco-leathers, eco-dyes, eco-skins). *Images* https://gup.unige.it/creative-food-cycles-experience

and co-management, with different levels of transversal interactions envisioning operational strategies dimensioned according to the expected impacts and policy frameworks, in order to define spaces of civic interactions and multi-functionality and daily care practices (places).

In this sense, to explore the "geographies of change" referred to the social food exchanges and the multi-performative food meta-changes of a new hyper-matter, appeals today to new models and agents (prosumers instead of consumers, new emerging start-ups instead of companies, living-labs and community flows).

Changes can be possible through new, more responsible social habits associated with the promotion and cognition of "food" processes, through new, more responsible social behaviours associated with the promotion and knowledge of "zero waste" processes and a new culture of recycling [23].

Designing urban spaces for conviviality and social rituals is, therefore, a design-led strategy to give voice to community interests through new fields of action, such as self-sufficiency, food sovereignty and cultural bio-diversity, creating narratives of urban–rural ties and experiences of circular economy.

The theme of the urban project and its relationship with eco-textiles may apparently seem far from the field of action of architects or urban designers and thus represents a horizon yet to be investigated, in order to combine the relationship between places of production, commercialisation and widespread sociality in the urban scenario.

Circular initiatives such as those described in the eco-textiles catalogue (Chap. 5) and new waves of designers show that the creation of materials from alternative sources is of great interest and can be a concrete means of reducing the environmental footprint of conventional materials [28], transforming food and agricultural waste and surpluses into valuable products and creating platforms, publications and databases, to share knowledge (Fabtextiles's books, DIYMaterials, Materiom, Food Waste explorer, Chemarts Cookbook).

In fact, this is a second level for design where eco-textiles can emerge as decisive materials in modern urban and architectural techniques and structures.

The application of textile structures in the field of architecture is, however, not entirely new. Of course, the use of fibrous elements for building enveloping spaces or roofs was already common in some primitive elements for elementary habitats, structures and shelters.

Much more sophisticated applications became relatively common from the 1960s/70s onwards with the development and refinement of synthetic and/or metallic fabrics. Projects such as those by Frei Otto for the German Pavilion at the Montreal Expo (1967) or by Frei Otto and Günter Benisch for the Olympic Park at the Munich Olympic Games (1972) were emblematic of an iconic moment associated with large textile structures (Figs. 4.11 and 4.12).

In recent times, some highly innovative projects designed by a new generation of architects are evidence of this. In fact, pioneering proposals such as the famous façade of the Allianz Arena stadium in Munich (Herzog and de Meuron 2005), the six textile eco-towers built over the Vallecas cogeneration plant (Soriano and Asociados 2010), the projects of the Lastra-Zorrilla studio (Palacio de congresos de Plasencia

Fig. 4.11 Frei Otto, Olympic Park at the Munich Olympic Games (1972). *Source* Jorge Royan. https://es.m.wikipedia.org/wiki/Archivo:Munich_-_Frei_Otto_Tensed_structures_-_5293.jpg

2013) or the Serpentine 2015 pavilion in London, a colourful and luminous structure covered by a textile membrane (Selgas-Cano 2015).

Other more examples are the ventilated façade for the new headquarters of Iguzzini in Barcelona, where the textile mesh is tensioned from the exterior structure (Josep Miàs 2018) or the Pavilion for the Yeosu Expo in Seoul (One Ocean 2012) with a responsive façade based on mobile modules composed of polymeric fibres, which are examples of reactive and dynamic properties that are currently in clear expansion.

We could even add the pavilion designed by Zaha Hadid Archs and ETH Zürich for the MUAC in Mexico, where knitting is used as a new textile technology—knitted in 3D—to create flexible surfaces for curved concrete structures, avoiding costly and time-consuming formwork (KnitCandela 2018) (Figs. 4.13 and 4.14).

The wide innovation of the world of eco-textiles, eco-leathers, eco-skins obtained as a result of processing food waste, packaging and/or eco-sustainable materials, has, however, recently gone beyond the dimension of design or fashion products to reach the urban scale, where experimentation with eco-fabrics and new spatialities has embraced the fields of architecture, installations and interior design, activating a virtuous cycle of actions aimed at qualitatively designing new meeting places and urban spaces.

Textiles are important materials in modern architectural technology, as they make it possible to construct lightweight structures with specific geometries. The way the fabric is produced, the density and thickness of the material play a significant role in architectural comfort. These characteristics influence the way we perceive textiles and their ability to distribute light and air, their acoustic, insulation and structural strength characteristics, etc. The production of textiles with a non-homogenous

Fig. 4.12 Textile contemporary architectural structures: eco-stacks in the energy plant of Vallecas, Madrid (Soriano and Asociados 2010), Palacio de congresos de Plasencia (Lastra-Zorrilla 2013), Serpentine Pavilion, London (Segas Cano 2015)

Fig. 4.13 Zaha Hadid Archs and ETH Zürich, Pavilion for the MUAC in Mexico, where 3D generated knitting is used as a new technology to create flexible concrete surfaces and structures (KnitCandela 2018)

composition, in which the highest density and strongest fibres are only used where necessary, significantly reduces the quantity of fibres and therefore the weight of the structure, without worsening its characteristics.

Therefore, the traditional method of textile production in large-scale architectural design involved shaping, stitching and welding, which are time-consuming processes because they consist of several production steps that significantly increase production time and costs.

Modern CNC knitting machines are capable of producing textured fabrics and complicated shapes with minimal human intervention. This allows complex, seamless three-dimensional shapes to be created quickly and without waste. However, the programming of knitting machines requires specific skills and experience, which is why, for example, in the fashion industry, where knitting is most common, this is carried out by knitting engineers who design textiles using a trial-and-error method.

But for large-scale customised architectural applications, this methodology significantly increases production time and costs, so new automated programming methods are required [29].

Research and experimentation in this field is therefore recently addressing these new challenges in prototyping, modelling and automating the production of knitted textiles for large-scale architectural applications in order to achieve innovative, more

Fig. 4.14 Spanish Pavilion at Expo Milan, 2015, by Federico Soriano and Dolores Palacios. The pavilion is a large-scale greenhouse, where crops are grown on artificial terraces that form a complete panorama, a perspective of the whole visible from the main entrance. Image courtesy of F. Soriano

performative and less impactful textiles using new sustainable materials and reducing waste.

A significant example of this is the "Senseknit" sensory pavilion, presented at MADE EXPO 2019, Design Week 2019 and the Tensinet Symposium 2019 in Milan, which combines tradition, innovation and digital culture, key elements in the transformation of the design and construction fields. The project is the result of an interdisciplinary research of the Politecnico di Milano conducted by the Material Balance research group, SAPERLab and TextilesHUB of the Department of Architecture, Built Environment and Construction Engineering and the Knitwear Laboratory of the Department of Design [30].

The pavilion aims to draw attention to the potential of textiles as a sustainable alternative to conventional building materials that can satisfy the aspects of safety and architectural comfort. Both can be controlled at the level of fibre composition and fabric production methods.

The aim of the "Senseknit" pavilion is to use engineered fibres and advanced digital knitting technology to investigate how textiles can influence architectural comfort while being both durable and safe.

The pavilion is constructed as a single curved wall, which bends and curves to form four partially closed areas, used to demonstrate the different architectural qualities of the textiles. The basic structure is made of CNC-cut wood and assembled into 22 individual panels that are easy to transport and assemble.

The entire structure is covered with 90 m2 of knitted textiles, optimised with digital knitting technology. This technology allows complex, seamless three-dimensional forms to be created quickly and without waste. Connected to performance-oriented modelling programmes, it allows for higher-performance fabrics using less material. In this way, patterns and compositions of heterogeneous materials can be achieved.

The four thematic areas of the pavilion offer different sensory comfort scenarios with specific performance characteristics: acoustic, climatic, visual and structural, thanks to the different types of textiles produced specifically for each area (Figs. 4.15 and 4.16).

The knitted fabrics that compose them, in fact, have been produced with performative fibres with advanced properties, in particular: a polyester fibre produced from the recovery of disused plastic bottles (produced by Newlife™ by Sinterama); special sound-absorbing fibres for the noise absorption effect (produced by HollowCore) and Lumen photosensitive fibres (produced by LineaPiù Italia) coated with photosensitive pigments, transparent to normal light, able to change colour when exposed to ultraviolet light.

The areas have been designed as a response to the needs of contemporary architectural space that evolves at a rapid pace. Acoustic comfort, for example, is an essential requirement for public, shared and crowded spaces. Fabrics for the acoustic part have been produced with a special sound-absorbing fibre and formed in a 3D pattern, which enhances acoustic performance.

The structural performance was improved by identifying the areas of the fabrics exposed to the greatest loads and stresses and using knitted patterns to provide greater load resistance. This results in a lighter structure while maintaining reinforcement only where necessary. From a climatic point of view, knitted fabrics, distributed in different densities, are used to control air movement and achieve a desired flow. From an optical point of view, the level of openness of the knitted fabrics, in different modes and intensities, helps to control the level of light and create visual effects.

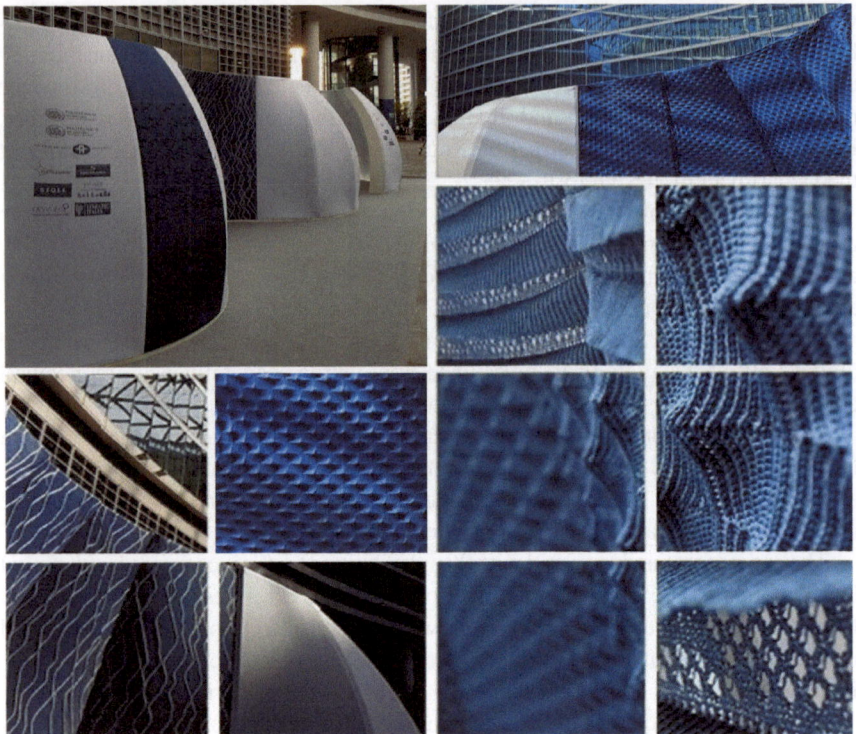

Fig. 4.15 Senseknit—a knitted sensorial pavilion, Material Balance Research, Laboratory, TextilesHUB, The knit lab, Laboratory SAPERLab. Picture by Elpiza Kolo. https://archinect.com/schools/project/5012532/senseknit-a-knitted-sensorial-pavilion/150145548

In 2021, at the 17th International Architecture Exhibition of the Venice Biennale, PoliMI's Material Balance Research group presented the MatRes project, an installation evoking the metaphor of a tree, understood as a living laboratory of resilient materials of biological, organic, sustainable and recycled origin.

The installation, structured as a cylindrical tree, houses within it a series of different biotextile "trunks" (each consisting of a composite material of natural origin such as mycelium, exhausted coffee, cellulose) with optical, acoustic or heat-sensitive properties of natural or recycled fibres, which react virtuously to stressful stimuli [31].

Paraphrasing the authors:

Working with resilient biomaterials calls for a subsequent transformation of spaces, places and relationships between the natural and built environment.

A new architecture will start from the chemistry and biology of materials, giving them voice and body, attending to their performances, attending to their functional potentials, in a synergistic relationship that today is exhausted: Designing with resilient biomaterials means assuming them as actors and protagonists with their forms, behavior, qualities, temporalities and possibilities.

Fig. 4.16 PoliMI's Material Balance Research Group *MatRes* project (17th Venice Biennale 2021)

Their origin, their study under the microscope, the prelevance of computational design, the life cycle of the used elements, begin a design model that is now available for a conscious and responsible future. In this new approach to matter and materials, choices will be guided by the environment, nature and life, which will intersect in a new relational logic, creating an increasingly integrated and holistic system [31].

Similar premises (combining parametric textile structures and reactive spatial responses) have led to other recent prototypes and installations, such as the one conceived by the IAAC in Barcelona (Institute for Advanced Architecture) for the FAM in Montpellier (Festival des Architectures Vivantes).

The IAAC installation, developed for the Festival des Architectures Vives 2014 in Montpellier, proposed to create an eco-textile construction (based on spun fibres obtained from the recovery of plastic for packaging) and generated by the smallest possible number of wood light-support elements, such as a meshed and tactile surface floating on a light and foldable structure. The pavilion also develops an interactive potential through a sensory experience that allows, by touching the fabric walls, to generate a soundscape. The textile structure—thanks to the combination of textile weave and sensors—reacts to the users by incorporating the senses of sight, touch and hearing, ultimately evoking a diverse reality. The recent installation by the Dutch studio Belén, which occupied the space of the Dutch Dubai pavilion at the World Expo Dubai 2020, with its "Curtain" project, a 44 × 14-m tent made of a completely recyclable plant-based PLA bio-fabric, also drew attention to the potential of using bio-fabrics, with a zero carbon footprint, in architectural applications.

The pavilion is designed as a closed-loop climate system where visitors can enjoy intense sensory experiences such as immersion in silence, heat, coolness, light, darkness, agriculture or water. In the arid climate of the Dubai desert, the pavilion is conceived as a temporary biotope, built with construction methods and biomaterials that make the concept of circularity comprehensible. A large tent made of bio-textile fibres structures the space, which has a central canopy. The tent and canopy are made from 100% bio-textiles, developed in collaboration with various associated research institutes. The tent is a large pleated fabric, 44 m wide and 14 m long, engraved using a laser technique. The PLA-based fabric material is made from renewable resources: indigenous vegetation, including date palms, moringa, mangroves and oleanders, form the basis of the knitted bio-fibres that make up the intricate laser-cut tent. The canopy is made from PLA, a sustainable alternative to polyester, and is highly UV resistant, offering greater durability in the sun than synthetic options. The exclusive use of locally sourced (and recyclable) materials ensures the lowest possible environmental impact and carbon footprint [32] (Figs. 4.17 and 4.18).

We have already commented on Neri Oxman's work with advanced bio-textiles.

In the field of performative installations understood as biosensitive prototypes, it is inevitable to mention her work with advanced bio-fabrics in the Silk Pavilion (2013), a silk thread structure created from 6500 live silkworms around a 3D-printed model, which explores the relationship between digital and biological construction, proposing methods that combine biological spinning and robotic weaving.

The project is based on the silkworm's ability to create a three-dimensional cocoon from a single thread of silk. The basic structure of the pavilion was created using

Fig. 4.17 Sense eco-textile structure (FAV-Montpellier 2014), courtesy of IAAC. www.iaac.net/project/unfold

26 polygonal panels of silk thread laid by a computer numerically controlled (CNC) machine, on which a flock of 6500 silkworms were placed.

Over three weeks, assisted by a robotic arm, they spun flat non-woven silk panels, locally reinforcing the gaps between the CNC-deposited silk fibres. Each silkworm spun a single silk filament about 1 km long, and together they produced a dome-shaped thread as long as the Silk Road.

This relationship between textiles, sustainability and technology, which is the basis of material ecology, is also reflected in Aguahoja's "Biopolymer pavilion" project (2019), which consists of two 3D-printed structures using 5740 fallen leaves, 6500 apple peels and 3135 shrimp shells as building materials that decompose naturally, enriching the environment rather than poisoning it. The result is a tissue-matter-organism of biopolymer compounds that sequesters carbon dioxide, enhances pollination, increases soil microorganisms and provides nutrients.

The matter and energy stored in the Aguahoja Pavilion will gradually give way to the natural process of decay and biodegradation, transforming into new biomass and ecologically stimulating plant growth.

The diversity of shapes and behaviours embodied in the Aguahoja structure reflects how nature constructs a wide range of multifunctional materials from a minimal set of molecular components, with no synthetic parallels. A material such as chitosan is found in the hard and tough exoskeletons of crustaceans such as shrimp and lobster, but also in the extremely thin and transparent membranes of dragonfly wings and in the soft tissues of fungi. Unlike steel and concrete, biocomposites made from organic materials are in constant dialogue with their ecological niches, able to adapt dynamically to changing environmental conditions. The project, based on the material ecology design approach, therefore focuses on the formation and degradation of materials (tissues, bioplastics, biomaterials) through design.

Fig. 4.18 Curtain and canopy for Dutch Pavilion at World Expo Dubai 2020, studio Belén. https://www.burobelen.com/projects/bio-textiles-world-expo-2020

As mentioned above, the installations and prototypes described so far, with more or less creative, performative or experimental radicalism, can be considered as "tests" of future or very near (technical-constructive) potentials, carried out on a larger scale on facades, interiors or surfaces or environmental control panels.

Natural fibres and eco-textiles are not only an alternative as aesthetic elements or building envelope systems, but also as building materials, in particular through the valorisation and transformation of textile waste into building materials.

In particular, natural fibres, such as jute, flax, coconut fibres, sisal and cotton, as reinforcements in building materials, are of great interest due to their advantages over synthetic materials, and one of their main advantages is their low environmental impact, low cost and wide range of applications [33] (Figs. 4.19 and 4.20).

It is well known that the "business as usual" scenario is not a viable option for a sustainable future and that different development models need to be identified. The construction industry, perhaps more than others, must reflect this urgency for change, as it is still plagued by a number of damaging factors, such as the use of high-impact materials, irreversible design solutions, inefficient processes and manufacturing.

The use of natural materials would trigger a different approach to construction, offering a number of benefits over traditional material options, such as lower CO_2 content, reduced health risks and lower costs.

As in the construction sector, the textile industry disposes of a significant amount of fibrous waste and post-consumer products, which is not only an environmental concern but also a waste of useful resources. Textile offcuts are often disposed of as waste products, which become an environmental problem due to their non-biodegradability, or by incineration, which releases highly toxic fumes.

Turning them into useful materials therefore has the dual function of eliminating waste and introducing a new product. Textile recycling can bring economic and environmental benefits by helping to reduce landfill space, the need to produce new materials and water pollution problems. Textile recycling involves the reuse of used clothing, fibre materials and production losses from the garment manufacturing process. The idea of using textile waste in applications that do not require a new industrial process is giving rise to a new research landscape [33].

In this sense, recent studies and experiments have explored the use of natural-based composites to create new sustainable building materials from the processing of natural and/or food fibres such as flax, hemp, jute, cotton, sugar cane and maize or from the recovery of textile fibre waste.

The PoliTO research group, in collaboration with a number of regional industries, has processed waste materials from local industrial supply chains—originating from the agro-food and textile industries (dry beans, chestnut bark, textile dust, maize husks)—to create an innovative natural insulation material, AGROTESs. The experiment aims to contribute to sustainable research by creating and testing a series of rigid, self-supporting and innovative panels for thermal-acoustic insulation in buildings [34].

Also in France, the architect Clarisse Merlet has patented an insulating, ecological, structural and aesthetic building material by recycling waste fabrics and producing a brick that can be used to make furniture, partitions or interior cladding (FabBRICK)

Fig. 4.19 Silk Pavilion I, Neri Oxman with Markus Kayser, Jared Laucks, Jorge Duro-Royo, Carlos David Gonzalez Uribe at MIT Media Lab, 2013. Images: https://oxman.com/projects/silk-pavilion-i

[35]. The use of textile fibres in specific applications can solve two problems: the elimination of an environmental pollutant and the provision of an alternative material for the construction industry [33].

Again from textile waste, the TeAM research group has developed a pre-feasibility study aimed at exploiting pre-consumer textile waste within the brick production

Fig. 4.20 Biopolymer pavilion, Aguahoja project, Neri Oxman + The Mediated Matter Group, 2019. Images: https://oxman.com/projects/aguahoja

chain, producing a clay brick containing polyester fibres (MAST), and carried out a pre-feasibility study aimed at exploiting pre-consumer textile waste in plaster mix mortars, with the aim of recycling wool and cashmere fibres (ReCash Plaster) [36].

We have already mentioned how elements with a strong architectural or urban component (facades, structures, panels) could soon be made using the beneficial properties of biomaterials and/or bio-fabrics or eco-textiles (Fig. 4.21).

Also in urban public spaces, interwoven meshes associated with fibres or yarns derived from food, plastic or mixed recycling can be used in various furniture or equipment, such as children's games based on ropes, fences, benches, reactive elements or sensorised surfaces, etc.

Schulberg Park has recently been updated (2011) with a clear innovative will and with the help of the ANNABAU artistic team. Designed in the shape of a pentagram, the park features a large climbing structure that surrounds a central landscaped area. Two green steel tubes form the backbone of this net structure made of strong recycled eco-fabrics, from which children can climb to vertical play platforms, combining art, play and nature.

Takino Suzuron Hillside National Government Park is home to one of the most amazing play structures in the world. The unique Takino Rainbow Net (2013) is the result of a collaboration between renowned designer Toshiko Horiuchi MacAdam and architecture studio Interplay Design and Manufacturing, Inc. This structure is made of 100% crocheted eco-textile bio-fibres that are interlaced and overlapped to create a light, colourful and fluffy environment that is as rich in experience as it is inspiring.

The Numen Designers' research-related experience: a metal net in an inflatable space with hermetically sealed floating black meshes (Yokohama 2021) gives way, during London Fashion Week (London 2021), to a labyrinthine circuit based on woven volumes made of biofabric meshes.

The Energy Carousel was one of the winning projects in a competition organised in 2010 by Carve, an Amsterdam-based group specialising in public playgrounds. Located on the Governeursplein (Dordrecht) and designed by the studio "Ecosistema Urbano", the installation promotes education through play, teaching children to generate alternative energy through their physical experience. The structure uses eco-efficient materials and a limited amount of steel in its structure, with a central body consisting of a tensegrity formed by strong textile ropes. The roof's colourful bio-fabric protects the children from rain and bright sunlight and features a swirling pattern to emphasise the "dynamic energy movements". LED lighting is used to minimise electricity consumption (Figs. 4.22 and 4.23).

Although the use of textile structures derived from waste or food surpluses is not always present in all these architectural or urban structures, it is implicit in the use of biodegradable eco-fibres. The improvement of processes and performances related to food by-products will allow, in the near future, a use similar to the examples described above, on a larger scale.

The proposed scenarios show the potential of digital knitwear and technical fibres as advanced technologies and materials that can revolutionise the way spaces are designed and inhabited. This intuitive environmental performance of knitted textiles

Fig. 4.21 FabBRICK, 16 tonnes of recycled textiles for the partnership with the JULES brand, for all their boutiques. https://www.fab-brick.com

Fig. 4.22 Numen Designers: A metal mesh in an inflatable net (Yokohama 2021) gives way, at London Fashion Week (2021), to a circuit based on braided volumes made of biofabric nets

with performative purposes transforms the ancient tradition of the past into a future perspective where the materiality of architecture will no longer be a fixed property, but a flexible condition linked to advanced bio- and techno- logical processes.

In the "age of entanglement", the multidisciplinary, or rather antidisciplinary, approach breaks down the boundaries between disciplines. Through engineering and chemical research, and through computational design, the possibility of understanding the "genetics and behaviour of materials" has opened up new matter programmes where biomaterials can be active generators of a new urban design based on new integrated systems between the old equation "material—machine—man—environment" [29].

For example, the tool of 4D printing, combined with the development of architectural software (Rhinoceros, Grasshopper and plug-ins for physical analysis, etc.), makes it possible to use materials whose properties can be programmed through the implementation of algorithms that simulate and reproduce behavioural models and parameters linked to the morphogenetic process, exploring new symbiotic relationships between material engineering, computational design and digital fabrication.

A new architecture will start from the chemistry and biology of these new responsible materials, following their performance and optimising their potential, reversing their functional and accessory relationships, thus defining a new design paradigm that promotes a homeostatic relationship with the environment, conserving resources and reducing waste.

Fig. 4.23 Toshiko Horiuchi MacAdam. Takino Rainbow Net (2013), a knitted playground made of 100% crocheted eco-textile fibres. http://www.knitjapan.co.uk/features/c_zone/horiuchi/work_p2.htm

References

1. López A, Alonso R (2020) Materiales: una historia sobre la evolución humana y los avances tecnológicos. Universidad de Burgos, Burgos. https://doi.org/10.36443/9788416283965
2. Gómez A (2021) Biotéxtiles, el auxilio de la biología para diseñar un futuro menos dañino, in Bio, May, 14th, 2021. https://gccviews.com/biotextiles-el-auxilio-de-la-biologia-para-disenar-un-futuro-menos-danino/
3. Ditty S (2019) How our clothes might change the future. In: Anyone//Anywhere (British Council). https://www.britishcouncil.org/anyone-anywhere/explore/digital-creativity/clothes-change-future
4. Caparrós R (2021) Biotextiles, el futuro de la moda, in El Escarabajo Verde, (October, 15th, 2021). https://www.rtve.es/television/20211015/biotextiles-futuro-moda/2187360.shtml
5. Materiales ed (2017) Informe sobre el mercado mundial de biotextiles, in Plataforma Tecnológica Sectores Manufactureros (November, 17h, 2017). https://www.platecma.com/inf orme-sobre-el-mercado-mundial-de-biotextiles/
6. Gausa M, Guallart V, Müller W, Morales J, Porras F, Soriano F (2003) The metapolis dictionary of advanced architecture. See the term *Naturartificial*. Actar Publishers, Barcelona
7. Oxman N (2019) Material ecology. In: IAAC BITS n. 10. Barcelona: IAAC ed, pp 127–133. https://iaac.net/iaac-bits-about/
8. AA.VV (2023) Fabricademy BCN—textile & technology academy: program & manifesto. https://iaac.net/educational-programmes/applied-research-programmes/fabricademy-textile-technology-academy/
9. Maurello ME (2018) Biotextiles: el cultivo de organismos vivos para fabricar telas in La Nación (February, 24th, 2018). https://www.lanacion.com.ar/moda-y-belleza/biotextiles-el-cul tivo-de-organismos-vivos-para-fabricar-telas-nid2111290/
10. Schröder J (2019) Circular design and the paradigm of Gestaltung in creative food cycles. In: Markopoulou A (ed) (2019) Responsive cities—disrupting through circular design. IAAC Institute of Advanced Architecture, Barcelona, pp 24–27
11. Marin J, De Meulder B (2018) Interpreting circularity. circular city representations concealing transition drivers. Sustainability 10(1310):1–24
12. Ellen MacArthur Foundation (2019) Cities and circular economy for food. Online at: http://www.ellenmacarthurfoundation.org/assets/downloads/CCEFF_Full-report-pages_May-2019_Web.pdf
13. Stahel W (2006) The performance economy. Palgrave Macmillan, New York
14. Schröder J (2020) Circular Design for the regenerative city: a spatial-digital paradigm, Hannover, Regionales Bauen und Siedlungsplanung Leibniz Universität Hannover. Book 1:17–31
15. Pericu S (2020) Food waste as a transitional key factor towards circular economy, Hannover, Regionales Bauen und Siedlungsplanung Leibniz Universität Hannover. Book 1:137–147
16. Ellen MacArthur Foundation (2014) The benefits of a circular economy. In: Crowther G, Gilman T (eds) Towards the circular economy. Accelerating the scale-up across global supply chains. World economic forum report. McKinsey & Company, New York. http://www3.weforum.org/docs/WEF_ENV_TowardsCircularEconomy_Report_2014.pdf
17. Fry T (2010) Design as politics. Berg Publishers, Oxford
18. Papanek V (2019) Design for the real world: human ecology and social change. Pantheon Books, New York
19. Escobar P (2018) Designs for the Pluriverse: radical interdependence, autonomy, and the making of worlds. In: New ecologies for the twenty-first century. Duke University Press, Durham
20. Mulgan G (2018) L'intelligenza collettiva ci farà riflettere. In: Morning future
21. Capra F (1999) Turn, turn, turn: understanding nature cycles, Liverpool Schumacher lectures. In: Capra F (ed) Ecoalfabeto, L'orto dei bambini. Viterbo, Stampa alternativa, p 52
22. Metcalfe R (2019) Food routes: growing Bananas in Iceland and other tales from the logistics of eating. MIT Press, Cambridge

23. Sommariva E (2020) Foodways: diasporic explorations at the age of (digital) discoveries, Hannover, Regionales Bauen und Siedlungsplanung Leibniz Universität Hannover. Book 1:159–171
24. Pollan M (2006) The Omnivore's Dilemma: a natural history of four meals. The Penguin Press, London
25. Newman D, Cepeda-Márquez R (2018) Global food waste management: an implementation guide for cities. World Biogas association. Sustainable Bankside Edition, London
26. Steel C (2009) Hungry City. How food shapes our lives. Random House, London
27. Schröder J, Haid C (eds) (2015) Food and the city. Regionales Bauen und Siedlungsplanung Leibniz Universität Hannover, Hannover
28. Camere S, Karana E (2018) Fabricating materials from living organisms: an emerging design practice. J Clean Prod 186:570–584
29. Anishchenko M (2019) Computational Knitting, an experimental design process for a performative textile system. https://www.materialbalance.polimi.it/portfolio_page/computational-kni tting-in-architecture/
30. Senseknit Pavillon, Politecnico di Milano (2019). https://www.materialbalance.polimi.it/por tfolio_page/senseknit-pavilion/
31. MatRes (2021) From a microscope to a prototype for a resilient architecture. https://www.mat erialbalance.polimi.it/portfolio_page/matres/
32. Buro Belén, Curtain and Canopy of Bio-Textiles for Dutch Pavilion at World Expo Dubai 2020, 2020–2022. https://www.burobelen.com/projects/bio-textiles-world-expo-2020
33. Pichardo PP, Martínez-Barrera G, Martínez-López M, Ureña-Núñez F, Ávila-Córdoba LI (2018) Waste and recycled textiles as reinforcements of building materials in natural and artificial fiber-reinforced composites as renewable sources. INTECH. https://doi.org/10.5772/intechopen.70620
34. Savio L, Pennacchio R, Patrucco A et al (2022) Natural fibre insulation materials: use of textile and agri-food waste in a circular economy perspective. Mater Circ Econ 4:6. https://doi.org/10.1007/s42824-021-00043-1
35. FabBRICK. https://www.fab-brick.com/fabbrick-english
36. Tedesco S, Montacchini E (2020) From textile waste to resource: a methodological approach of research and experimentation. Sustainability 12(24):10667. https://doi.org/10.3390/su1224 10667

Chapter 5
Knitting-Food: A Global Catalogue of Eco-textiles

5.1 International Catalogue of Innovative Eco-textiles

This chapter is structured as a catalogue of experiments and prototypes in the field of eco-textiles—and eco-products derived from them—with a strong innovative character.

Although in the previous chapters we talked about bio-textiles and bio-materials, in this chapter we have decided to extend the research to eco-textiles, i.e. those textiles whose production and characteristics aim to make the product more sustainable, thus reducing its environmental impact compared to products in the same category.

In order to better understand the ambiguity of products defined as ecological, it is first necessary to see what is meant by the term "environmental impact", defined by the EMAS Regulation (Eco-Management and Audit Scheme, i.e. a voluntary instrument created by the European Community to assess and improve the environmental performance of organisations and to provide the public with information on environmental management) as "any change to the environment, whether negative or positive, resulting in whole or in part from an organisation's activities, products or services". It follows from this definition that achieving a low environmental impact product means ensuring that the product has a reduced contribution to environmental change throughout its life cycle (which includes the extraction of raw materials to the final disposal of the product).

Therefore, in this catalogue the illustrated results, generated by the fruitful combination of design, science and sustainable industry, show numerous examples of new alternative processes, as well as concrete, economically and functionally effective products, some of which have already been promoted and commercialised as R&D examples in pioneering companies of national and international renown.

The catalogue proposed in these pages synthesises this research, as an operational and functional manual, generated around the use, reuse and recycling of agricultural products, understanding the concept of Food-Matter not only as Eat-Matter, but also as Hyper-Matter, search matter and research material.

In the division and classification of this reasoned catalogue, the food industry itself has served as a basis, through the definition of a series of specific sectors, called to group different types of food according to their characteristics: these include the meat sector, the fishing sector (with fish processing), the fruit and vegetables sector (with the cereals sector), the dairy and milk sector, mushrooms and bacteria, and sectors related to the food industry such as food packaging and other natural materials.

"Food surpluses, food waste, food scraps, food leftovers and food packaging remnants are some of the most important (re)cycled resources involved in this set of proposals. Obviously, other important elements derived from today's ability to work with living or inert microorganisms (bacteria, mycelium, cells, algae, etc.) can be processed, reinformed and embedded in the design of new, evolving and responsive eco-textile materials (eco-tissues, eco-fabrics, eco-skins and eco-dyes) obtained through innovative processes from different "food" items.

As we pointed out in Chap. 2, the possibility of reconciling ethics and social responsibility in the fields of innovative design has opened the doors to new products derived from fruits, vegetables, milk, mushrooms, coffee, shellfish, algae and many others related to the (re)use of food waste/discards, which can be classified according to this criterion.

A and B. Fruits and Vegetables

Banana // Orange //Pineapple // Coconut // Grapes and Wine // Coffee // Mango //

Fruit Fibres // Avocado // Artichoke // Beetroot // Vegetable Fibres

C and D. Animal by-products and Fishing sector

Milk // Animal by-products // Fishes and Shellfishes // Algae

E. Mushrooms and Bacteria

Mycelium // Bacteria

F. Plants

Plants

G. Packaging and Other

Plastics // Gums // Paper and Carton // Hair // Various

Table 5.1 summarises the classifications of the different food and non-food categories and subcategories of the food industry that make up the Global Eco-Textile Catalogue. This international catalogue summarises around **69 eco-textiles** through project sheets and is a basic reference in the great experimentation that is developing in this sector.

The type of product obtained from waste materials is indicated in the last column of Table 5.1:

- Eco-fabrics and eco-tissues
- Eco-leathers and eco-skins
- Eco-dyes and eco-pigments
- Eco-film and eco-membrane.

Table 5.1 Classifications of the different food and non-food categories and subcategories of the food industry that make up the Global Eco-Textile Catalogue. Author: G. Tucci

A	Fruits	1	Banana	Bananatex	Eco-fabric
		2	Orange	Orange fibre	Eco-fabric
		3	Pineapple	Piñatex Sustrato	Eco-leather Eco-fabric/film
		4	Coconut	Malai	Eco-leather
		5	Grapes	Tejido Conectivo	Eco-leather/film
		6	Coffee	Etimo TômTex	Eco-leather Eco-leather
		7	Mango	Puur Allegorie Fruit leather	Eco-leather Eco-leather Eco-leather
		8	Fruit fibres	Neflium Tejido Conectivo Revitalising Yarn Kaiku	Eco-leather Eco-leather/film Eco-fabric Eco-dyes
B	Vegetables	1	Avocado	Natural Dye Club	Eco-dyes
		2	Artichoke	Boertex	Eco-leather/film
		3	Beetroot	Morphling	Eco-leather
		4	Vegetables Fibres	Kaiku Dyelicious	Eco-dyes Eco-dyes
C	Animal by-products	1	Milk	Duedilatte Caseina	Eco-fabric Eco-film
		2	Animal by-products	Mestic Inner Values The Meat Factory Gold Feathered Fabrics	Eco-fabric Eco-film Eco-dyes/leather Eco-film Eco-fabric
D	Fishing sector	1	Fish and shellfish	Fish Skin Crabyon TômTex	Eco-leather Eco-fabric Eco-leather/film
		2	Algae	Algae Sneaker Line Algear Kelp Carbo Lichen Weaving Water	Eco-leather Eco-film/fabric Eco-leather Eco-leather Eco-fabric Eco-fabric
E	Mushrooms and bacteria	1	Mycelium	Ephea Fungi Narratives Mylo The Pure Hyphae Project MuSkin We Grew Together Eco Warrior	Eco-leather Eco-fabric Eco-leather Eco-leather/film Eco-leather Eco-leather Eco-leather

(continued)

Table 5.1 (continued)

		2	Bacteria	reGrow	Eco-leather
				Kombucha Couture	Eco-leather
				Scoby-compo	Eco-leather
				A Baby, A Beast	Eco-leather
				Moving Pigments	Eco-dyes
				Maqui Biotextile	Eco-leather
	F. Plants	1	Plants	Eco Warrior	Eco-leather
				InterWoven	Eco-fabric
				Maqui Biotextile	Eco-leather
				Lovr	Eco-leather
				Climafibre	Eco-fabric
				Flocus	Eco-fabric
				Latex	Eco-leather
				Bio-Invasive Textile	Eco-fabric
				Library	Eco-fabric
				Juhla	Eco-leather
				Mader	Eco-fabric
				Fique	Eco-fabric
				Simbiosis	Eco-leather
				Yerma	Eco-leather
				Desserto	
G	Packaging and other	1	Plastic	Rifò	Eco-fabric
				Plastex	Eco-fabric
				Plastigela	Eco-leather
		2	Gum	Gomma	Eco-fabric
				Gum-Tec	Eco-leather
		3	Paper and carton	Consumption of Heritage	Eco-fabric
				Abitinuovi	Eco-fabric
		4	Hair	Human Material	Eco-fabric
				Loop	Eco-fabric
				Contemporary	Eco-fabric
				Hairwork	Eco-fabric
				Weaving Water	
				Wolfwall	
		5	Various	Made In	Eco-fabric

5.2 From Fruits to (A1./A8.)

Banana // Orange // Pineapple // Coconut //
Grapes and Wine // Coffee // Mango // Fruit Fibres

A1	Banana	Bananatex	Eco-fabric
A2	Orange	Orange Fibre	Eco-fabric
A3	Pineapple	Piñatex Sustrato	Eco-leather Eco-fabric/film
A4	Coconut	Malai	Eco-leather
A5	Grapes and Wine	Tejido Conectivo	Eco-leather/film
A6	Coffee	Etimo TômTex	Eco-leather Eco-leather
A7	Mango	Puur Allegorie Fruit Leather	Eco-leather Eco-leather Eco-leather
A8	Fruit Fibres	Neflium Tejido Conectivo Revitalising Yarn Kaiku	Eco-leather Eco-leather Eco-fabric Eco-dyes

In the field of new eco-textiles made from fruit scraps and waste or surplus, we can first highlight the group of eco-textiles made with organic elements based on banana (Bananatex), orange (Orange Fibre), pineapple (Sustrato) or fruit fibres (Revitalising Yarn).

We can also add eco-leathers based on pineapple (Piñatex), coconut (Malai), grapes and wine (Tejido Conectivo), coffee (Etimo, Tôm-Tex). Mango (Puur, Allegorie, Fruit Leather) or from diversified fruit fibres (Neflium), capable also of generating eco-dyes (Kaiku) that compose new varnish alternatives (Fig. 5.1).

Fig. 5.1   iStock CC, by MEDITERRANEAN

A1. | From BANANA to

Bananatex®
QWSTION, Zurich, Switzerland, 2018
#banana #circulareconomy #sustainability #eco-fabric
www.bananatex.info, www.qwstion.com

FRAMEWORK:
The non-fruit bearing species of banana, abacá (Musa Textilis), has been grown for centuries for use as a textile fibre. Also known as "Manila hemp", banana fibre production and its trade has centred primarily around The Philippines due to the plant's abundance and quick regrowth. Bananas grow on plants, rather than trees, with the fruit technically being a berry. The leaf sheath around the base of this herbaceous flowering plant is where you find the hidden fibre resource—not in the fruit at all. The abacá variety grows non-edible fruit, but some farmers can also take the banana fibre from a plantain species, therefore utilising the whole plant.

DESCRIPTION:
Bananatex® is the world's first durable, technical, biodegradable and plastic-free fabric made purely from abacá banana plants. Cultivated in the Philippine mid- and highlands within a natural ecosystem of sustainable mixed agriculture and forestry, the plant requires no chemical treatments, pesticides, fertiliser or extra water. Its self-sufficiency has made it an important contributor to reforestation in areas of former rainforest that were once cleared for logging, whilst enhancing biodiversity and the economic prosperity of local farmers. Bananatex® was developed by Swiss bag brand and material innovators QWSTION in collaboration with a yarn specialist and a weaving partner in Taiwan with the aim to cause positive change.

As an open source project Bananatex® offers a truly circular alternative to the synthetic fabrics that dominate the market today. Since its launch in October 2018 the company has won a variety of international awards such as the Green Product Award 2019, the Design Prize Switzerland Award 2019/20, the German Sustainability Award Design 2021, as well as the PETA Fashion Award 2023 for «Innovation of the Year».

PRODUCTION:
The fabric is incredibly strong and durable, while remaining soft, lightweight and supple. The natural white colour reflects the real colour of the fibres and is not dyed. An all-natural wax coating can be applied to make the fabric waterproof. The resulting fabric has a smooth and distinctive hand-feel. In December 2021 the Bananatex® fabric was Cradle to Cradle Certified® Gold, the most advanced standard globally for products that are safe, circular and responsibly made.

Photo credit: © Yves Bachmann

A2. | *From ORANGE to*

Orange Fibre
ORANGE FIBER SRL, Catania, Italy, 2014
#orange #circulareconomy #sustainability #eco-fabric
www.orangefiber.it

FRAMEWORK:
In Italy, every year, the citrus industry produces about 700,000 tons of citrus *"pastazzo"* whose waste disposal, done with methods that are not always correct, represents a considerable cost for the citrus production chain and for the environment. Residue of citrus fruit, including pulp, oil and peel, has been largely processed by industries to generate substances for other food, cosmetics or feedstuff. Orange Fibre is committed to creating circular good practices throughout the textile-fashion supply chain reducing waste as well as pollution by recycling bio- and subproducts that would otherwise be disposed with excessive costs for both the industry and the environment.

DESCRIPTION:
Orange Fibre is the first and leading company in the world to produce sustainable fabrics from bio-products of the citrus transforming industry through an innovative process patented in 2014 in Italy and subsequently extended to different countries around the world. The process is based in the transformation of citrus juice past (the so-called "pastazzo") into a creative, low-impact and innovative fibre. It starts from the subproduct of the citrus industry, that is, from all that remains after the production of juice and which otherwise would have to be discarded with economic and environmental costs. Through a fully traceable and transparent supply chain, the firm transforms these subproducts into the perfect ingredient for brands and designers who care about sustainability. The elements made with these processes are biodegradable: through a special composting system they are capable of be decomposed in an ecological way.

PRODUCTION:
The technology is based on the extraction of cellulose suitable for spinning from the subproducts of the citrus industry, which represent 60% of the weight of the whole fruit and which would otherwise have to be disposed of. Thanks to their patented process, this cellulose is recovered and transformed into textile fibre. The citrus textile results 100% soft and silky hand feel, lightweight, opaque or shiny. The fibre is then extracted from the cellulose which is initially white and is dyed with natural dyes, an important alternative to the use of synthetic, polluting and unsustainable dyes. Furthermore, through sophisticated nanotechnologies for the firm, it is possible to enrich the fabric with microcapsules containing citrus essential oils which are gradually released onto the skin, hydrating it; manufacturing a high-quality material and contributing to reshape the sartorial experience, the project links two pillars of Italian heritage—textiles and food—responding to demands of both innovation and sustainability in the fashion industry.

A3. | *From PINEAPPLE to*

Piñatex®
ANANAS ANAM, United Kingdom-Philippines, 2013
#pineapple #circulareconomy #sustainability #eco-leather
www.ananas-anam.com

FRAMEWORK:
Natural fibre-based composites are more and more under intensive study due to their eco-friendly nature and peculiar properties. Pineapple leaves fibre is one of the abundantly available wastes materials of Philippines (2.198 million tonnes in 2018) but with an unexpressed reuse potential supportive to farmers, textile sector, environment and market demands. The wide range of available natural fibres, as important agricultural biomass, can reduce the pressure on tropical forest and woodlands, which normally produce 30–40% waste materials. For this reason, pineapple leaves fibre can also be implemented in value added products and creative processing, due to no additional environmental impact on the territories.

DESCRIPTION:
Piñatex is an innovative plant-based textile made from pineapple leaf fibre, developed over seven years of R&D by Dr Carmen Hijosa. The company works to the values of a circular economy, supporting rural communities in a combination of research and innovation to enhance a sustainably sourced textile, created with low water use and no harmful chemicals or animal products. Shocked by the environmental impact of mass production of leather and chemical tanning, the project looks for sustainable alternatives to PVC. Its goal is to connect ecology and economy to create a new industry that is socially and environmentally responsible. Piñatex is made from up to 95% bio-based and renewable materials, bringing new income streams to partner farms and farming groups who can efficiently use the waste of their crops. This creates a textile non-woven suitable for use in fashion, accessories and upholstery that has been used by many brands worldwide.

PRODUCTION:
After the pineapple has been harvested, the remaining suitable leaves are collected in bundles and the long fibres are extracted using semi-automatic machines. An innovative automated decorticating machine assists this process, allowing farmers to use greater quantities of their wasted leaves. The fibres are extracted, washed and then dried naturally in the sun, or during the rainy season in drying ovens. The dry fibres go through a purification process to remove any impurities that result in a lint-like material. This fluff-like pineapple leaf fibre (PALF) is blended with a corn-based polylactic acid (PLA) and undergoes a mechanical process to create *Piñafelt*, a non-woven web that forms the basis of most *Piñatex* collections. The finished textile is distributed to designers, who can use it as a sustainable alternative to leather and synthetic materials in footwear and fashion industry.

A3. | *From PINEAPPLE to*

Sustrato
Andrea De la Peña, Ciudad de Mexico, 2019
#pineapple #circulareconomy #sustainability #eco-fabric #eco-film
www.futurematerialsbank.com/material/pineapple-waste/

FRAMEWORK:
Sustrato is a project that explores the application of ancestral techniques to transform the pineapple industry into four different biomaterials and sustainable products (felt, agglomerated, bioplastic and ropes). The pineapple plant leaves are considered the main residue (around one ton each day), which represent 75% of the harvested product. They are usually thrown away through garbage trucks or piled up in their production fields for degradation. This unregulated practice generates the production of greenhouse gas emissions, a change in the soil PH, an increase in pests that affect nearby crops and even promotes infectious diseases that can affect workers. The materials and products developed offer sustainable and compostable alternatives with competitive properties and prices, compared to similar products currently on the market.

DESCRIPTION:
The felt and agglomerated material are made from extracted pineapple leaf fibre. They have demonstrated great properties for sound reduction and rebound. Hence, some products for acoustic absorption were designed with each material because they can reduce more that 10db, while similar products on the market only reduce about 40db. The rope is also made of pineapple leaf fibre and it has high strength. It is softer and lighter than similar natural ropes on the market. The bioplastic is made of pineapple leaf bagasse and natural binders, that is permeable translucent material with a degradation time that takes no longer than two years, which makes it an ideal material for solid product packaging.

PRODUCTION:
The pineapple leaves are collected and then boiled to facilitate fibre extraction, which is carried out through a decortication machine. The fibre is separated from the bagasse to be used in the fabrication process of different materials. To make felt and the agglomerated material, the fibre is carded and manually organised on a net fabric to form a layer, which is felted on a rigid surface with an artisanal technique. To make the pineapple fibre rope it is only needed to separate the fibre and place it in Sustrato's braiding machine. Then, when the fibre is braided and stretched out a small amount of the diluted natural binders is sprayed on it and finally the resulting rope is dried. For the bioplastic, the bagasse is mixed and boiled with water and natural binders to get a paste. The paste is poured on a smooth mould and dried until the material separates itself from the mould surface. Then, the resulting layer is hung for a couple of hours to let it dry completely.

A4. | From COCONUT to

Malai
Malai, Kerala, India, 2019
#Coconut #bio-leather #circulareconomy #sustainability #eco-leather
https://malai.eco/

FRAMEWORK:
Malai is born with the desire to produce new bio-leather processes, reconciling a pure, simple philosophical approach to manufacturing with a sophisticated understanding of environmental science and technological processes resulting in a product that is unique.

DESCRIPTION:
Malai is a newly developed bio-composite material made from entirely organic and sustainable bacterial cellulose, grown on agricultural waste sourced from the coconut industry in Southern India. It is a flexible and durable with a feel comparable to leather or paper. It is water resistant and because it contains absolutely no artificial "nasties" it will not cause any allergies, intolerances or illness. It is a completely vegan product and as such you could even eat it.

PRODUCTION:
Malai materials are produced through various artisan techniques combined with innovative technologies. A large stock of products and materials, in optimal conditions of qualitative finishing, make up the pieces of the collection. As a complementary sideline, *Malai Fin(it)e* is the name of a collection of commercial accessories launched with a successful crowdfunding campaign in 2019. All the pieces of the collection are made with materials supported by a coconut-based substrate, and are biodegradable and, as such, compostable at the end of their useful life. Users often believe that the longevity of a material is decisive in its product quality. However, considering another perspective, a certain duration time can be considered for a material and its application as a fashion accessory. Fashion is always subject to change and taste and as such, the longevity of the material may not always be an advantage when it comes to end of life. *Malai Fin(it)e* pieces are designed with circularity in mind: pieces with a minimal, yet temporal and functional aesthetic.

A5./A8. | *From WINE and FRUIT FIBRES to*

Tejido Conectivo
Studio Elia Gasparolo, Mendoza-Argentina, 2020
#fruitandwineskins #winetissues #fruitissues #circulareconomy #eco-leather #eco-film
www.eliagasparolo.com

FRAMEWORK:

Connective Tissues: perfumes, textures and flavours can today be turned into skin and tissues that sustain, connect and give new meaning to experiences and cycles. Shells and remains of what is consumed can become a second skin, transforming the past into a new life. A future that makes an offering to memory and links old and new ties and origins; trusting in past experiences and their impact on the future implies giving new meaning not only to the materials but also to history itself.

DESCRIPTION:

Designing dresses, necklaces or collections of accessories, from fruit peels (oranges, tangerines, etc.), remains of wine and lavender—three elements with emotional memory that connect the materials with their roots—is the basic concept of *Tejido Conectivo, Biotextiles con Memoria* (Connective Tissue, Biotextiles with Memory) the name of a bio-textile line that emerges like a second skin, intense, sensory and deep.

PRODUCTION:

In this sense, the products generated are based on various combinations or "recipes".

- The *Malbec Wine Petal Collier* (*Malbec Necklace*) is a Bio textile based on a combination between Malbec wine and corn starch with other ingredients: cloves, cornstarch, water, vinegar and vegetable glycerine.
- With the remains of the Malbec wine it is possible to generate the *Lavender and Wine Necklace* another bio tissue based on corn starch wine and lavender combined with remains of linden tea, corn starch, water, vinegar and vegetable glycerine.
- With citrus peels such as tangerines it is possible to design the fabrics of a dress or a kind of leather for necklaces. These are significant materials from which emanate certain perfumes and flavours with a certain importance in memory and places of origin.
- Other materials allow more flexible applications and characteristics similar to the body that are used for a new type of contemporary jewellery.

Tejido Conectivo is a work that allows the participation of the entire family environment, near and far. The experimentation with these and other technologies makes it possible, at the same time, to discover the current scope of biomaterials and bio-textiles or bio-tissues and bio-skins and propose an open-source system to share formulas and recipes with the rest of the community of researchers, creators or designers.

A6. | *From COFFEE to*

Etimo
Etimo Lab—Camila Castro Grinstein/Buenos Aires, Argentina, 2017
#foodwastematerials #circulareconomy #sustainability #eco-leather
https://etimobiomateriales.com/

FRAMEWORK:
New collaborative alliances, interdisciplinary projects and artistic crossings make up the plot of a new sensitivity and a new specialty, related to biomaterials and the functional use of organic and food waste. A sign of the times where a banana peel is as relevant as a highly complex digital system. Sowing design, cultivating technology and harvesting bio-manufactured products: a future that is "planted" as a timely and possible challenge.

DESCRIPTION:
Etimo is a research project linked to the production of biomaterials generated from the recycling of organic and biodegradable waste. Directed by Camila Castro Grinstein, it works in a transdisciplinary way, dialoguing with biology, chemistry, and design to create new elements and textures. New containers and packaging elements or bowls are some of the pieces developed from coffee compounds, cabbage and eggshells or other "seasonal residues". Etimo also focuses on the transmission and dissemination of knowledge through various projects, including a series of workshops and self-learning and awareness-raising kits, with a strong emphasis on the circular economy and innovation.

PRODUCTION:
Among the main elements produced by Etimo it is worth highlighting:

1. Met Textiles: a leather-like material made from discarded grass for the manufacture of natural dyes. Collaborating in a circular way with certain dye factories, the raw material used for the biomaterial was reused after being used as a natural dye for the textiles themselves. The vegetable residue is thus reused twice, giving it great value as a natural resource.
2. Be Fish. Material generated for the Be.Fish jewellery brand as part of its sustainable packaging. Designed from coffee to provide a sensory experience when receiving the pieces and rings of jewellery with the aroma of coffee, generating a strong contrast between the dark colour of the container and the shine of the metal. Once used, the packaging material can be composted and/or disposed of on the ground to fertilise the plants.
3. Etimo Pack Kit: a kit designed to be able to take an online biomaterials course at home. All the elements contained in the ETIMO box (box and mould in turn, designed in 3D printing and made with PLA biomaterial) as well as the bags of the various elements provided (jellies, agar, glycerine, anti-mould products) are made of biodegradable starch.

A7. | *From MANGO to*

Puur

Àmber van de Ven, Amsterdam, The Netherlands, 2021
#fruitwaste #mango #circulareconomy #sustainability #eco-leather
www.futurematerialsbank.com/material/mango/

FRAMEWORK:

The PUUR project is a material exploration project centred around the creation of a fruit leather.

The designer created "recipes" that are open source, so that others can make their own with biomaterials. A new research question was formulated: How can we implement biomaterials in people their everyday life in an easy way?

DESCRIPTION:

This resulted in the creation of a fruit leather bag. With the use of leftover mango's from the markets and the waste of curtain fabrics, the designer has created a strong material that has the feel and look of leather. Using mango leather can be used as an alternative to animal leather and vegan leather, which is often "leather" made out of plastics.

The making process is based on a recipe that is used to make healthy candy for kids; fruitleather rolls. With only the blended pulp of the mango, it is possible to make a piece of leather if you spread it on a layer of fabric. After that, it needs to dry in the oven till all the water is evaporated.

PRODUCTION:

PUUR is a project in which a bag is made of mango leather. The mangoes were collected from a local supermarket where they were no longer good enough for sale. The mangoes are pulverised and spread over old textile layers. After a long drying process, it has become a strong material that is reminiscent of leather.

By turning this biomaterial into a recognisable everyday product, such as a bag, people become acquainted with the possibilities and applications of this material in an accessible way.

A7. | *From MANGO to*

Allegorie
Koen Meerkerk and Hugo de Boon, Rotterdam, Netherlands, 2021
#fruitwaste #fruitleather #circulareconomy #sustainability #eco-leather
https://allegoriedesign.com/pages/ourmaterials

FRAMEWORK:
In 2018, global production of mangoes was 55.4 million tons, led by India with 39% of the world's total. China and Thailand were the next largest producers. In 2017, world mango imports totalled $2.8 billion, with the U.S. (23.2%), Netherland (9.9%) and Germany (7%) being the largest importers. The fruit is inedible before ripening and turns squishy very fast, so harvesting and transportation is highly time sensitive. It is estimated that more than 30% of mangos are wasted during harvesting and transportation globally, with some regions experiencing much higher waste rate (for example, post-harvest losses for the West Africa are 50–80%). In the U.S., approximately 20% of mangos in the country's supermarkets are discarded due to defects, over ripeness and many times just not good-looking enough to be picked by consumers, per a study the USDA conducted in 2016. In Australia, mango joins the 3.1 million tons of edible food that is thrown away every year by Australian households, according to the National Food Waste Strategy. And all these numbers only include mangoes discarded in supermarkets, and do not include the mangoes that are already lost prior to arriving at supermarkets or after being purchased.

DESCRIPTION:
The goal is to utilise the fibres and components of mango waste, such as peels and seeds, to create a material that mimics the texture and appearance of leather. Rich in natural fibre even in over ripeness, mangoes are a great source for making elastic and durable materials.

PRODUCTION:
Through partnerships, the discarded mangoes are collected from supermarkets, shred them down into "smoothies" and turn the entire fruit into sheets through an eco-friendly process. With the help of natural additives as biding agent, fortified by coating and backing, the juicy fruit is transformed into strong leather-like material that's ready to be shaped and conditioned. The amount produced depends on the seasonality of the fruit, and how many mangoes are thrown away by importers, supermarkets and other stores. Ensure that the waste is clean and free from contaminants. The process of transforming mango waste into leather-like products involves several steps.

A7. | *From MANGO to*

Fruit Leather
Koen Meerkerk and Hugo de Boon, Rotterdam, Netherlands, 2021
#fruitwaste #mango #circulareconomy #sustainability #eco-leather
www.fruitleather.nl

FRAMEWORK:
The fashion industry is one of the most environmentally destructive industries in the world. It is responsible for 10% of all global carbon emissions. Every year, more than a billion animals are slaughtered for leather production. The cleaning process of these materials alone produces approximately 650 million kilos of CO_2.

In this context, the project considers what would happen if we did not see fruit waste as waste, but as a valuable starting material for possible alternative systems.

Regarding food waste and its relationship with increasingly scarce resources, every year 1300 million tons of food are thrown away worldwide: approximately one third of all world food production. 45% of all fruit produced for consumption is thrown away.

30% of the agricultural land on earth is used to produce food that will eventually be wasted and 40% of the food harvested remains in the fields, because it does not meet supermarket standards. 10% of all greenhouse gases emitted in developed countries are used to produce food that will never be eaten.

DESCRIPTION:
Fruit Leather is a project conceived by a team of young designers—Koen Meerkerk and Hugo de Boon—who have decided to reduce the waste of fruit and vegetables—in particular the rests of handle or mango—and reduce, at the same time, the costs for their disposal. To do this, they collect organic waste from the markets and transform it into material to make bags, furniture and clothes. Specially the mango fibres permit to obtain a vegan leather-like material that is then sold to designers all over the world.

The aim of the project is to raise awareness of the problem of food waste and demonstrate that a solution exists. The vision here at Fruitleather Rotterdam is not only to spread awareness of the food waste issue, but also to show how waste in general can be used in a positive way.

PRODUCTION:
Over the years, the team has experimented with bio-natural mixed models aimed at turning wasted fruit into a leather-like material through various synthesis processes. The products obtained make it possible to introduce ecological and animal-friendly products to the market.

Fruit Leather seeks to create a variable and versatile material that can be turned into footwear, fashion accessories, upholstery, furniture and other possibilities that are still open.

A8. | *From FRUIT FIBRES to*

Neflium

Studio Hole—Bertín López, Ciudad de Mexico, Mexico, 2020
#rambutan #sustainability #biotextile #circulareconomy #eco-leather
https://studiohole.com/main

FRAMEWORK:

Neflium is a biomaterial developed by Studio Hole as part of a research initiative focusing on the utilisation of food waste, employing a bio-upcycling philosophy developed in collaboration with Biology Studio and applying various other circular methodologies. It can be categorised as fruit leather, crafted from organic waste sourced from rambutan and other biodegradable components. While bearing some aesthetic resemblance to animal leather, Neflium boasts distinct and unique properties.

Depending on the formulation and components, it can vary in flexibility and durability, while consistently retaining a sweet aroma and a textured surface that is both rough yet intriguingly natural. The most important importance aspect of its composition is the exclusive use of biodegradable and non-toxic ingredients, designed with a "Cradle to Cradle" approach, with a keen focus on the natural cycles of organic matter.

Currently, Neflium is available in solid black as well as a warm colour palette ranging from brown to red, all derived from food waste sources.

DESCRIPTION:

Like leather, Neflium can be used to craft objects with traditional leatherwork techniques or can even be craft them with digital fabrication tools. Some explorations have already been undertaken utilising both methods. The currently specific applications designed by the studio, are made with the same eco design principles of Neflium, for example, avoiding the use of glue, paint, toxic chemicals, or dangerous components.

PRODUCTION:

Many everyday items are traditionally crafted from animal leather, a process that consumes significant amounts of water and involves the use of toxic chemicals, resulting in substantial environmental damage. In response to this, our aim was to design a collection of objects to replace commonly used disposable products made from toxic and non-biodegradable materials. These materials not only harm humans but also negatively impact other forms of life and their interactions. The central goal of the project is to demonstrate that what some consider waste can be repurposed to create meaningful objects that eventually reintegrate into the natural cycle without harming ecosystems. Neflium is a noble material crafted with a circular approach with clear applications in fashion and design, that proposes an alternative to traditional leather and addressing the environmental concerns associated with its production.

A8. | *From FRUIT FIBRES to*

Revitalising Yarn
Aleandra Gil La Rocca, Barcelona, 2021
#fruitwaste #revitalizingyarn #circulareconomy #eco-fabric
https://www.futurematerialsbank.com/material/fruit/

FRAMEWORK:
Revitalising Yarn is made from nuts, fruit waste and algae, offers energetic properties, an alternative to the textile world, and at the same time, it starts creating awareness in human minds and social consumption. The Textile industry is known as the second-largest polluter in the world, not only because of the unsustainable materials we use but also because of the process, there's always new matter, even with the organic.

DESCRIPTION:
The creation of a Revitalising Yarn allows you to fly, a vast universe of ideas on how to work together with nature and not against it, therefore a new more conscious world will come, a world full of possibilities.

PRODUCTION:
Thread made out of the waste of Pistachio, Peanut, Walnut Shells and Avocado Seeds.
    Revitalising Yarn is made with a completely artisan process, beginning with the simplicity of drying the stone or shell of the fruit. Making it a powder is a must so it is able to blend with the other organic ingredients and become a paste that can then be extruded by hand and turned into this "pasta" like shape. The process will be done after letting it dry for a few days in the sunlight.

A9./B4. | From FRUIT FIBRES and VEGETABLES FIBRES to

Kaiku
Nicole Stjernswärd, London, United Kingdom, 2019
#foodwaste #colordyeing #recycling #circulareconomy #eco-dyes
https://kaiku.bio/, www.stjernsward.co/kaiku-living-color

FRAMEWORK:
Currently most synthetic pigments are toxic or made of ambiguous materials where the intense colour appears, generally, as an artificial "contamination" of products obtained through the principles of the circular economy. Originally, the pigments were derived from nature, such as the blues of lapis lazuli stones, the yellows of ochre clay and the reds of the flattened wings of beetles. Vegetables such as onion were traditionally used to dye cloth. These methods have gone out of fashion with industrialisation and the introduction of cheaper pigments derived from petrochemicals. But the effect on people and the environment is often disastrous. Paints can release petrochemicals into the air long after they have dried, causing respiratory problems and damaging the ozone layer. Industrial effluents containing synthetic dyes seep into the water system, poisoning aquatic life and posing a great danger to human health.

DESCRIPTION:
Kaiku offers an alternative system that uses food waste that would otherwise rot in landfills to produce non-toxic pigments. It is a system that turns plants into powdered paint pigments using vaporisation technology. Avocados, pomegranates, beetroots, lemons and onions are just some of the fruits and vegetables that can be placed into the Kaiku processes and turned into the raw material for paints, inks and dyes. It offers an alternative system that uses food waste that would otherwise rot in landfill to produce non-toxic pigments. Kaiku means echo in Finnish: Kaiku colours sometimes express themselves or behave in surprising ways, reminding you that they came from living plants.

PRODUCTION:
The skins and husks from food waste are boiled in water to produce a dye, which is transferred to a Kaiku system's own repository. Together with hot air and in a pressurised environment, this dye is forced through an atomising nozzle into a glass vacuum chamber. The fine mist produced is hot enough to vaporise almost instantly, and the dry particles are drawn through the chamber into the collection bin. Because the pigments are dry powder, this means they can be used as an additive in almost any type of paint. Regarding materials, the bioplastics associated with agar, bacterial cellulose, paper, cloth, plaster and wood veneer are bases that allow highly efficient applications. Because they contain tannins, avocado skins and peels produce a ruby-red dye that appears as orange as paint or dyes fabrics a pinkish colour. Pomegranates and onions also produce a more yellow tint. Adding vinegar or baking soda to the stain is one way to modify the resulting colours.

5.3 From Vegetables to (B1./B4.)

Avocado // Artichoke //Beetroot // Vegetables Fibres

B1	Avocado	Natural Dye Club	Eco-dyes
B2	Artichoke	Boertex	Eco-leather/film
B3	Beetroot	Morphling	Eco-leather
B4	Vegetables Fibres	Kaiku Dyelicious	Eco-dyes Eco-dyes

In the field of new eco-fabrics and/or bio-tissues made from various vegetables, we can mention the group of eco-dyes made with organic elements based on avocado (Natural Dye Club) or other mixed vegetables and fibres (Kaiku, Dyelicious).

We can also add eco-leathers and films obtained from artichoke (Boertex) or beetroot (Morphling) (Fig. 5.2).

Fig. 5.2   Pixabay CC, by Serpae

B1. | *From AVOCADO to*

Natural Dye Club
Rebecca Desnos, UK, 2023
#avocado #sustainability #pigments #circulareconomy #eco-dyes
https://rebeccadesnos.com

FRAMEWORK:
Avocado dye has a history going back generations in South America, and it's most definitely a natural dye. Natural dyeing helps cultivate a sense of connection to nature, whether you're foraging for plants in your kitchen (avocados!) or gathering pine cones in the woods. Natural dyeing is a calming and meditative craft and is a wonderful way to engage in a creative and mindful practice.

DESCRIPTION:
Natural Dye Club is a new way to learn how to dye with plants and to "fully understand the dye processes with bite-size videos that clearly and simply teach you the why, not just the how. Easily adapt the recipes and use different dyes and materials. Stay inspired with a fresh, new dye project each month, plus you'll have access to the library of past videos and recipes to dip into whenever you like".

PRODUCTION:
Following, an example of the basic dye process. It is completely possible to experiment by pre-treating fabrics, changing the quantities of water and ingredients, the time spent heating the dye, or by continuing to heat the dye after the fabrics or yarns are added.

1. Put eight avocado stones into a saucepan containing 2 L of water (increase according to the amount of fabric you want to dye);
2. Bring saucepan to boil, then leave to simmer. The colour should start to change within half an hour;
3. After dye has turned a deep enough colour, heat can be turned down (I did this after 2 h);
4. Remove the avocado stones and debris, leaving nothing but the dye behind—this could help achieve a more even colour as any fabric touching the stones could potentially end up a different colour;
5. Add fabrics or yarns to the dye (you could continue heating, but I chose to try it with the heat off);
6. Leave for several hours, stirring every now and then until you are happy with the colour (I left my samples for between 9 and 16 h);
7. Remove fabrics or yarns with tongs, rinse (with a gentle soap optional) and leave to dry.

B2. | *From ARTICHOKE to*

Boertex
Rebecca van Caem, Amsterdam, The Netherlands, 2021
#artichoke #sustainability #fabriacademy #circulareconomy #eco-leather #eco-film
https://waag.org/en/artichoke

FRAMEWORK:
Rebecca van Caem, the designer of Boertex, pioneers a unique approach that combines fashion design with culinary art, exploring the intersection of fashion and sustainability by repurposing seasonal food waste in collaboration with local suppliers such as Dassemus vineyard and Lindehoff. By repurposing vegetable waste and promoting circular design principles, Boertex contributes to the development of eco-textiles and circular economy practices. The experimentation with the artichoke fibres, in combination with a cellulose-based recipe, marks the beginning of Boertex.

DESCRIPTION:
Boertex represents a biomaterial garment crafted from artichoke fibres, extracted from an abundance of seasonal vegetable waste. The project embodies the journey from food waste to fashion design, showcasing the innovative potential of biomaterials. Boertex possesses versatile properties, allowing it to be bent, laser-cut and ultimately biodegradable.

PRODUCTION:
While Boertex garments are currently not wearable, they exhibit remarkable properties such as water repellency. However, prolonged exposure to water will result in degradation, underscoring the ongoing development of the biomaterial. Boertex serves as a proof of concept, offering a glimpse into the future of sustainable fashion. The potential applications of Boertex extend beyond clothing, with possibilities ranging from furniture upholstery to interior design elements. One of the unique advantages of Boertex lies in its variability, as each batch of biomaterial reflects the unique characteristics of the food waste utilised, resulting in one-of-a-kind pieces that defy replication. Additionally, during the research process, Rebecca developed a biomaterial cookbook, which is open-sourced to share knowledge and empower others to work with sustainable biomaterials.

B3. | *From BEETROOT to*

Morphling
Katrijn Westland, The Netherlands, 2021
#beetroot #vegetables #biomaterials #circulareconomy #eco-leather
https://www.futurematerialsbank.com/material/beetroot/

FRAMEWORK:
Morphling is an interactive installation covered by a biodegradable "lick-able" skin that breathes, smells, moves and ages. The main ingredient of Morphling's skin is five hundred beetroots, harvested from Dutch farmlands.

DESCRIPTION:
After the harvest, the beetroots are sliced and dried for several hours. Once the drying process is finished, the beetroots are ground into a fine powder. Together with the rest of the ingredients, the powder is mixed into a paste. Then, layer by layer, the paste is thinly spread onto a fabric canvas. When dry, these layers can be pulled from the canvas and Morphling's skin is formed.

PRODUCTION:
If the idea of "beetroot fabric" has gained traction, it might involve extracting fibres or utilising components of beetroot in a process like other plant-based fabrics. Here's a general outline of how a hypothetical process for beetroot fabric work:

1. Extraction of Fibres: Beetroot fibres could be extracted from the plant's stalks, leaves, or other parts using a process like retting or other fibre extraction methods.
2. Processing and Spinning: The extracted fibres undergo processing to remove impurities and then spun into yarn or thread. This can involve mechanical or chemical processes to achieve the desired characteristics.
3. Weaving or Knitting: The spun beetroot yarns are woven or knitted into fabric. The choice of weaving or knitting depends on the intended use of the final textile product.
4. Dyeing and Finishing: If colour is desired, the fabric may undergo dyeing using natural or synthetic dyes. Finishing processes can be applied to enhance the fabric's softness, durability, or other desired properties.
5. Quality Control: The produced fabric undergoes testing for various properties, such as strength, colourfastness and overall quality, to ensure it meets industry standards.
6. Manufacturing End Products: The beetroot fabric is then used to create various textile products, ranging from clothing to home textiles.

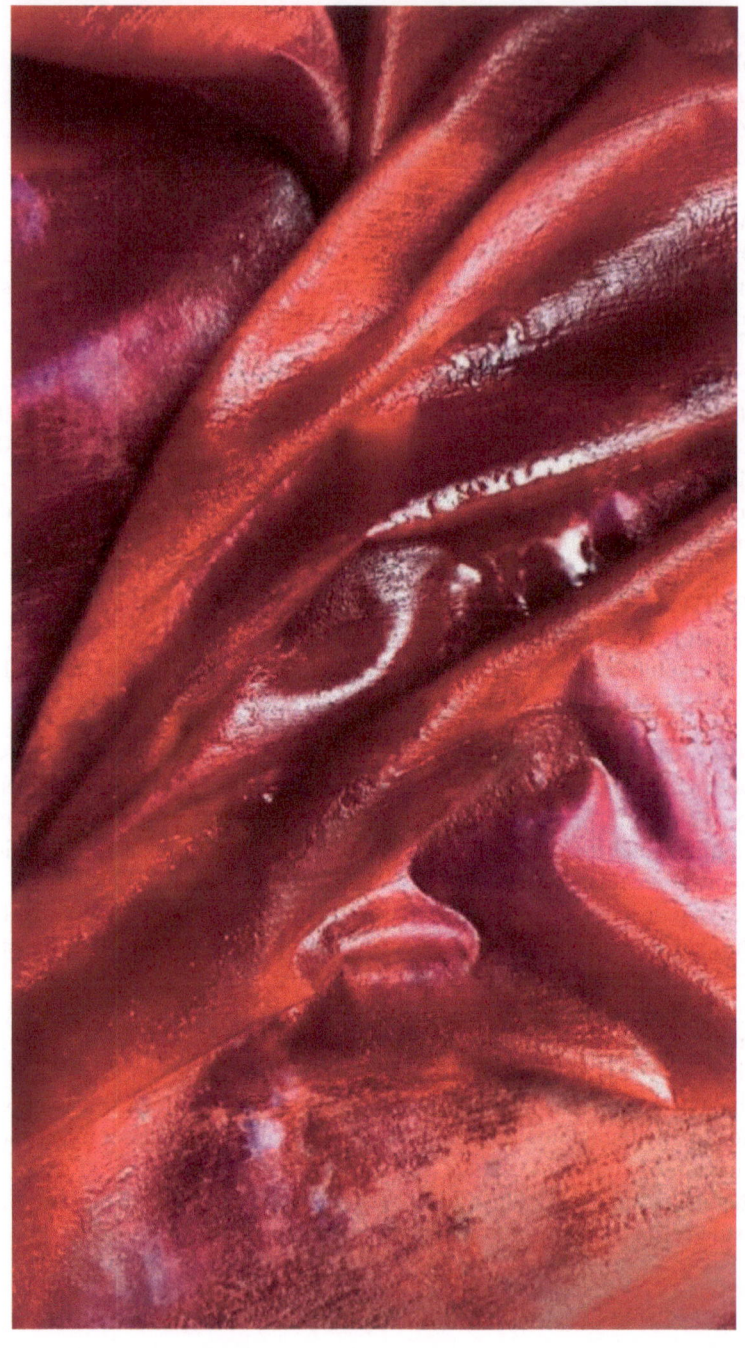

B4. | *From VEGETABLES FIBRES to*

Dyelicious
Eric Cheung and Winnie Ngai/Hong Kong, China, 2012
#fruitwaste #recycling #circulareconomy #sustainability #eco-dyes
https://www.dyelicious.hk

FRAMEWORK:
Textile dyeing is the second largest polluter of water in the world. Though it seems unrealistic to completely replace the highly cost-effective chemical dyes with natural ones derived from plants, it's time to think about the individual actions that we can take to lessen this impact on the environment with the aim of driving sustainable consumption. Choosing naturally dyed garments is always a good start. In Hong Kong there are many natural dye workshops where speed is everything, offering a valuable opportunity for city dwellers to get a closer look at the true, beautiful colours of plants. At the same time, large quantities of food end up in landfills every day in an affluent city like Hong Kong. To explore the possibilities of natural dyeing in the city permits new alternative approaches to turning food waste into dye for a self-sufficient innovative operation.

DESCRIPTION:
Dyelicious is a Hong Kong start-up that uses kitchen waste to make dyes that can decompose naturally and do not yield any pollution. The project turns food waste into high-quality clothing and other products through a process known as natural food dyeing. The products decompose naturally and do not yield any pollution, unlike the typical factory garments that may emit toxins into rivers and oceans. Dyelicious has worked with notable brands seeking to lower their carbon footprints, sourcing waste from Zara, Adidas, Towngas, Starbucks and Calbee. The company also works with schools to substitute their chemical paints with non-toxic products and hosts workshops for families at retailers across Hong Kong.

PRODUCTION:
Natural food-dye uses a series of processes that include extraction, liquid preparation and colouring. In order to up the quality of the dye, additional mordants are included so that, different hues can be transformed, the colour sharpness can be improved, and even different colours can be created. The team goes to the garbage dumps to pick up leftovers, buys discarded and crooked fruits and vegetables at the markets, recovers coffee grounds in cafes, and combined traditional vegetable dyeing techniques to turn these vegetable waste into dyes. Natural plant wastes are used as dyes: cloths using natural dyes are brightly coloured, and because the raw materials used are not chemical products, they are safer and non-toxic. Dyelicious insists on using 100% natural materials even though natural dyes have a hard time binding to fabrics. The team spent six years experimenting how to get a variety of colours from vegetable waste collected from wet markets and supermarkets.

5.4   From Animal By-products to (C1./C2.)

Milk // Animal by-products

C1	Milk	Duedilatte	Eco-fabric
		Caseina	Eco-film
C2	Animal by-products	Mestic	Eco-fabric
		Inner Values	Eco-film
		The Meat Factory	Eco-dyes/leather
		Gold	Eco-film
		Feathered Fabrics	Eco-fabric

In the field of new eco-fabrics or bio-tissues made from milk and milk products, several results can be highlighted (such as Duedilatte, Caseina).

This is a sector of growing importance that is extending its potential to other clothing accessories such as buttons, necklaces, pins, etc., very close to the bioplastic by-products derived from casein.

In the field of new eco-fabrics or biotissues derived from the meat sector, we can cite various experiences linked to animal by-products (Mestic, Feathered Fabrics). We can mention the group of eco-dyes produced with meat elements (The Meat Factory) or interesting eco-leathers (The Meat Factory) or films (Inner Values, Gold) linked to meat waste (Fig. 5.3).

Fig. 5.3 Pixabay CC

C1. | *From MILK to*

Duedilatte
Duedilatte—Antonella Bellina, Pisa, 2013
#milk #circulareconomy #sustainability #naturalfiber #eco-fabric
www.duedilatte.ch

FRAMEWORK:
The milk market in Italy has about 30 million tons of waste every year. Milk is a too precious raw material to be wasted and this is why the firm *Duedilatte* aims to enhance the production surpluses of the agri-food chain by transforming them into a new resource in the sustainable textile sector.

Furthermore, with the research and development team, they devote also resources to becoming a point of reference in the techno-sustainable textile sector, creating new innovative fibres starting from agri-food surpluses such as coffee yarn and rice yarn.

DESCRIPTION:
Since its foundation in Pisa in 2013, *Duedilatte* has been manufacturing innovative yarn and textile fabrics in Italy, starting from the amino-acids protein derived from casein extracted from milk. Thanks to their professional multi-disciplinary team of engineers, spinners, weavers and marketing experts, they work to create a product with innovative properties. Milk yarn is naturally antibacterial, thermoregulatory and gives the fabric softness and silkiness. *Duedilatte* yarns and tissues are completely natural, respect the environment and have extraordinary qualities: the fibre is obtained with circular processes that can enhance the industrial surpluses of the agri-food sector.

Thanks to the collaboration of diversified industrial partners, *Duedilatte* has created a vast range of yarns and knitted tissues, ideal for the production of clothing, furnishings or to be used in the para-pharmaceutical and automotive sectors.

PRODUCTION:
The milk that is used was previously discarded, but through this type of recycling, it can have a second life. Casein, a noble milk protein, is separated from the whey and subsequently isolated and denatured. From this the amino acids are extracted which, combined with an innovative viscose-based spinnable solution, are transformed into a textile fibre.

The new fibre is subsequently spun, and the yarn thus obtained is transformed into tissue. The tissue is purged of its raw materials with a detergent-free wash and finished (dried) ready in its most classic appearance: milky white colour.

The generated fibre is antibacterial, the derived fabric is very soft, transpiring and thermoregulating, it has a bright appearance and is silky to the touch.

C1. | From MILK to

Caseina

Universidad Diego Portales and Margarita Talep, Santiago de Chile, Chile, 2015
#milk #casein #circular economy #sustainability #eco-film
www.margaritatalep.com/Caseina-desarrollo

FRAMEWORK:

Casein is a precise and detailed research on biopolymers related to food products recycling. The project focuses on the possibilities of the generated material, taking it to different manufacturing and application scenarios. The project started as an academic investigation in the Sustainable Processes and Products Workshop of the UDP (Diego Portales University) led by the designer Margarita Talep. In the workshop, the investigation and the project are prosecuted so that a formal proposal is reached in a reduced time.

DESCRIPTION:

Caseina is, in fact, a bioplastic made from bovine milk. The project explores the methods of processing protein (casein) extracted from cow's milk as a natural alternative to oil-based polymers, using the milk with other biopolymers to produce objects. The obtained material is 100% biodegradable and constitute a new sensorial experience.

PRODUCTION:

The proposal was initially based, in effect, on obtaining a set of vessels to contain, for a short time, cold, fresh or dry food, and eat instantly. Subsequently, the project has given way to design processes with the most experimental material, generated in a much more free and imaginative way, producing various jugs, skins, vessels and decorative elements. One of the properties of casein is to intensify colours and enhance lights, so these elements enhance this property. Different morphologies, textures and colour palettes are developed in each piece.

C2. | *From ANIMAL-BY PRODUCTS to*

Mestic®

Jalila Essaïdi, Eindhoven, The Netherlands, 2016

#manure #circulareconomy #sustainability #newnaturalfabric #eco-fabric

https://jalilaessaidi.com/cowmanure/

FRAMEWORK:

Mestic® is a natural fabric made by recovering cattle manure. The objective is to reduce the environmental impact of intensive farming which is one of the main causes of air pollution, if not global warming. This is demonstrated by a study published in the Applied and environmental microbiology journal which analyses the process of decomposition of manure, declaring it responsible for the emission of nitrogen oxides, or greenhouse gases harmful to the planet, especially if produced in excess. It is on the basis of these discoveries that the Dutch designer Jalila Essaïdi decided to propose a solution capable of reducing the impact of cattle farming without compromising their production capacity.

DESCRIPTION:

With Mestic® the manure is saved and transformed into a biofabric which thus becomes part of the ever-growing number of natural yarns used in the fashion sector. Mestic® is the result of an innovative process created by Jalila Essaïdi, which starts from the cellulose contained within the manure to create new eco-friendly materials such as plastic, paper and natural fabrics.

PRODUCTION:

Following a sustainable manufacturing process, Mestic® is obtained by dividing the liquid parts from the solid parts of the manure; from the latter, cellulose is obtained which is in turn processed with the help of substances extracted from the manure in a liquid state; the compound derived from the previous phases is finally transformed into cellulose diacetate, a liquid polymer from which it is possible to create fibres, paper and plastic in total respect of nature.

Manure is gold: we conceive it as waste that is destroying nature but in manure there are many nutrients, chemical agents, phosphate and cellulose that we can reuse. Presented during a fashion show of clothes designed by the Dutch designer and subsequently declared one of the winners of the H&M Foundation's Global change award, Mestic® is yet another example of how waste can become a new business opportunity.

Picture and Credit: Fashion show Mestic® Jalila Essaidi—Clothes from Cow Manure—Photo Credit by Ruud Ba

C2. | *From ANIMAL-BY PRODUCTS to*

Inner Values
Tobias Trubenbacher, Germany, 2018
#cattleintestine #circulareconomy #sustainability #pigbladder #eco-film
https://tobiastruebenbacher.com/ux-portfolio/inner_values/

FRAMEWORK:
Due to livestock farming and industrial slaughter methods, the prices of animal products have decreased enormously compared to the recent past. While just some decades ago farm animals were highly valued and mostly all of their resources were further processed, in present times only the tastiest and easiest to prepare parts of an animal are used. Today less than half of an animal is really further processed in Germany. All the rest goes to animal rendering industrial plants and more directly to landfill. In addition, since consumers are rarely in contact with the raising and slaughtering process of animals, they started to be disgusted by the so-called by-products of animals. Tobias Trübenbacher initiated his "Inner Values" project: two seating pieces of furniture out of tanned and further processed cattle intestines and pigs' bladders, transformed them into soft leather. Thereby, the former poor reputation of the supposed "waste products" is being replaced and infused with opposite values.

DESCRIPTION:
Inner Values reveals that these skins can have equal qualities as conventional leather after they were cleaned, pickled, greased and tanned for several weeks. The outcomes demonstrate that these materials can have equal qualities as conventional leather. In addition, the pieces of furniture also stimulate us to rethink the handling of animal resources in our society and to question our unreflected waste culture.

PRODUCTION:
Trübenbacher started the project with various material studies in order to explore animal skins, to understand their characteristics and qualities and to try out different possibilities to conserve them. The materials were experienced with processes such as drying, stretching, weaving, prepared with glue and resin or blown up. The designer also experimented with both traditional and new tanning processes. Based on the findings of these first trials, the designer decided to focus on pigs' bladders and cattle intestines, which are both waste materials, usually thrown away after slaughter. Firstly, the innards are cleaned with a vinegar solution and freed of fat or meat residues. After washing the purified materials in water they are tanned for several weeks with natural tannins. During this process, the skins, which mainly consist of proteins, these proteins contract and the fibres solidify and merge together. Next, the designer regressed the bladders and intestines by repeatedly massaging a mixture of train oil, Vaseline, curd soap and tallow in the material. Afterwards, the materials were kneaded, stretched and tumbled in order to transform the stiff, brittle and wrinkled skins into light, soft leather, which was then stuffed with recycled cotton wool, sewn shut and attached to a seat frame.

C2. | *From ANIMAL-BY PRODUCTS to*

The Meat Factory
Shahar Livne, Eindhoven, the Netherlands, 2014
#bloodink #circulareconomy #sustainability #eco-dyes #eco-leather
https://www.shaharlivnedesign.com/the-meat-factory

FRAMEWORK:
In worst-case scenarios, slaughterhouses, dump blood and other leftover matter from animals into sewers and waterways, and in best-case scenarios, reuse it for animal feed, fertilisers, food colourant and iron supplements. Shahar Livne has developed a new handmade "bio-skin" using materials derived from slaughterhouse waste and by-product streams in the Netherlands and using wasted blood as a colourant and a plasticiser. The "Meat Factory" project is a collection of biomaterials with the aim of creating dissonance between attraction and disgust, natural and industrial attitudes, focusing on blood as a material and colour on all its meanings and symbolism.

DESCRIPTION:
The project results are displayed in two directions: 1. The use of blood as a pigment produces an exciting range of natural colours in screen-printing techniques. 2. Bioleather is produced with waste from the meat industry and low-value materials, such as boon, fat and skin trimmings as a suggestion to create new valid values in a polluting and often cruel industry that treats animals as products. As part of the Meat Factory project, designer Shahar Livne investigated the construction and deconstruction of living subjects and the wasteful and cynical treatment of animals and natural resources by humans, inspired by the philosophy of the "Nose to Tail" attitude that involves the use of the whole animal also using discarded blood as a pigment and plasticiser for the deep red material, and finally designed a pair of sneakers with alternative leather inserts made from the fat, bones and blood of animals taken from the waste streams of Dutch slaughterhouses in new material.

PRODUCTION:
The project explores the emotional and psychological connection with nature that has changed since food and materials have been industrialised and consumers have been alienated from the source, content and use of animal-based materials. While the main part of sneakers is made of recycled local Nappa leather, each shoe has a dark red panel in the centre of a material that the designer describes as bio leather, as a nod to the material's potential as an alternative to leather, while the soles of the sneakers are made of recycled rubber and the insoles are made from recycled cork. Livne hopes to develop her bio-leather material to the point where it could be used to make the entire shoe and other leather products. Collection and Preparation, Mixing and Straining (To remove any impurities or clots, strain the mixture through a fine mesh or cloth), Testing and Adjusting the proportions of blood, binding agent and air dry the pigment for future uses.

C2. | *From ANIMAL-BY PRODUCTS to*

Gold
WINT Design Lab, Germany, 2021
#collagen #sustainability #bioplastic #circulareconomy #eco-film
https://www.wintdesignlab.de/gold

FRAMEWORK:
The global demand for synthetic textiles is steadily increasing, while their production continues to consume large quantities of petrochemicals, energy and non-circular sourced raw materials. GOLD is a design research project exploring the properties and potentials of Goldbeater's skin, the outermost tissue layer of the cow gut. This extremely thin, elastic, tear-resistant and air-tight membrane was historically used as a separating layer when beating leaf gold. This research investigated its outstanding properties, determined the tissue structurally and biochemically, and sequenced its RNA. Its goal is to synthetically replicate the native material properties to develop a biopolymer and produce high-performance textiles.

DESCRIPTION:
Through interdisciplinary artistic research, WINT Design Lab explored the aesthetic and functional design features of the native tissue. Specifically prototyping its material properties in the field of today's outdoor garments. While experimenting, WINT researched forgotten archives and old scientific papers for descriptive process steps. Collagen is the most abundant biopolymer in nature, therefore animal skins have been used by humans throughout history. For example, the Inuit made water-repellent gut parkas from similar tissue types to keep hunters dry in their kayaks. Following processes and production steps WINT applied advanced methods of fabrication including robotic yarn laying and material lamination. The experiments resulted in a taxonomy of process steps that were documented in videos and material samples. Based on these findings, WINT designed a new generation of water-repellent outdoor jackets to demonstrate the future application field of a biosynthetic collagen-based textile.

PRODUCTION:
Within the research, a variety of process prototypes and samples ranging from robotic yarn lying to mono-material lamination and sewing techniques have been explored. At the collagen research institute FILK, the native tissue was extracted, as a by-product from a local butcher. The wet native tissue was frozen and sent to WINT Design Lab. There was dried on cotton fabric. After drying it was laminated with bone glue (collagen glue) to build multi-layer compounds. These techniques were explored by hand as well as with a UE5 co-bot. For the robotic yarn-laying, a new end-effector was developed, allowing to precisely laminate yarn on the tissue itself. The yarn was then laminated into a sandwich compound to function as a stiffening material. The robotic yarn laying approach relied on parametric design methods.

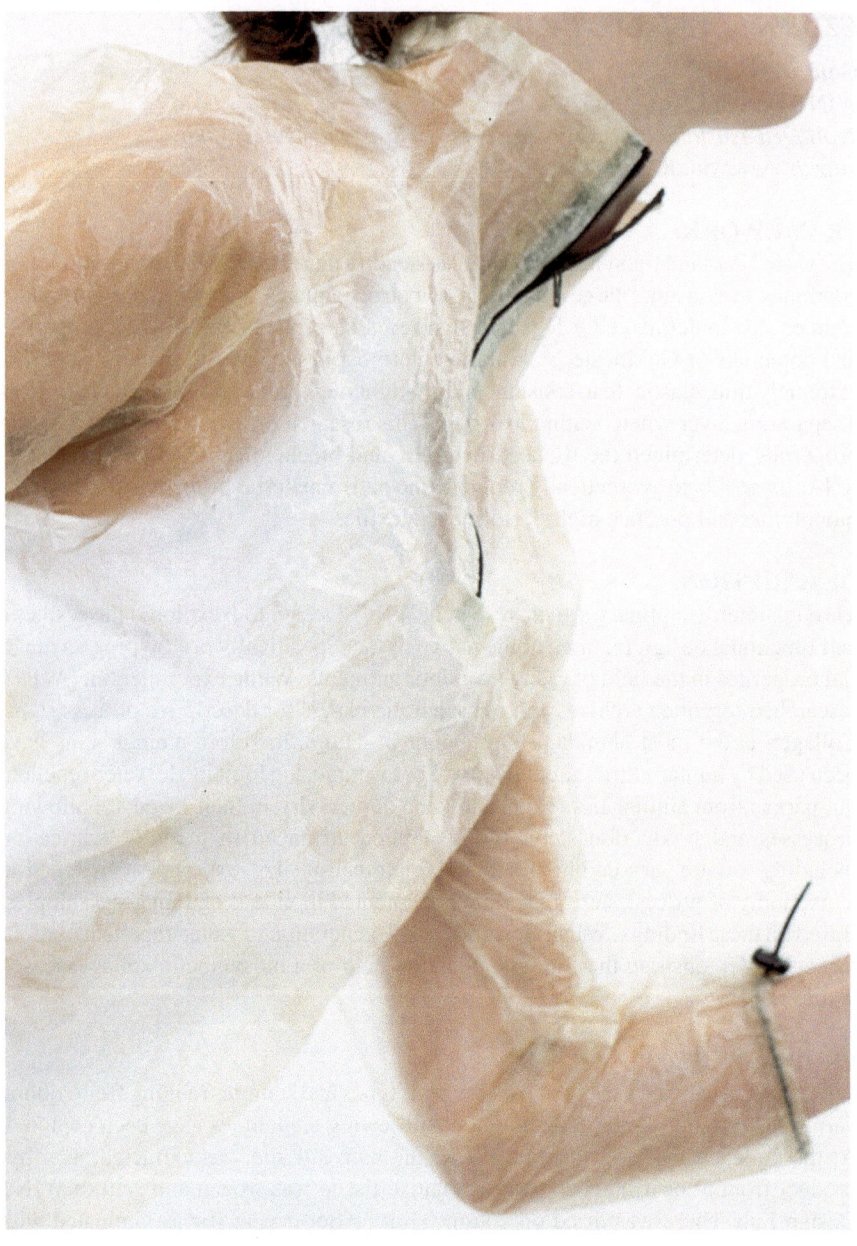

C2. | *From ANIMAL BY-PRODUCTS to*

Feathered Fabrics
Pascale Theron Studio, South Africa—The Netherlands, 2018
#ostrichfeathers #sustainability #reuse #circulareconomy #eco-fabric
www.pascaletheron.com

FRAMEWORK:
The ostrich feather was once a highly valued commodity during the 19th Century, as Victorian and Edwardian women sought out big plumes to decorate their flamboyant hats. Since then, the feather has fallen from grace and now its main use is removing dust or in carnival costumes. Pascale Theron Studio aim to enhance the value of the ostrich feather, a very specialised product coming from the 150-year-old ostrich feather industry in Oudtshoorn, South Africa. Going beyond the aesthetic and decorative and returning it to its former glory in a more integrated and practical way, within a modern society where big, feathered hats are no longer stylish.

DESCRIPTION:
Handwoven textiles made from ostrich feathers. The feathers themselves are woven into the warp while creating this very special textile. In the ostrich farming industry, the animal's main purpose is to be slaughtered after one year, for their nutrient-rich meat. After spending only, a couple of weeks in Oudtshoorn, South Africa talking to farmers and locals, the issue was clear that there was a need for change within the ostrich farming industry. Collecting feathers from the ostrich is the removal of dead material and is the equivalent of cutting fingernails, causing no pain to the bird. It could be a very functional interior textile, as well as a new solution to the fashion industry, as the new "fur", without the slaughter of animals, associated with fur. Introducing an alternative industry where the ostriches are only kept for their feathers, could mean that they could live their full lifespan while providing feathers without deterioration in quality for up to 35–40 years.

PRODUCTION:
Only the natural uncoloured feathers are chosen, as they would be found on the bird. By removing the central shaft, and only using the soft, thread-like barbules, the feathers can be combined to make a yarn that is then woven into a fabric. The aim of the project is to bring the manufacturing back to Oudtshoorn and have the feathered fabrics locally produced in the place where it all began. Using the feathers within the fabric differs from the way that ostrich feathers are currently being used in the fashion industry (usually just sewn or glued on top of the fabric as decoration). The breathable, washable, soft, warm and incredibly lightweight textural quality of the feathers mean they can be used in a variety of practical ways. By harnessing the quality of the feathers and creating a new craft through these feathered textiles, it could not only save many animals' lives but also could create a new industry of economic value in the small town of Oudtshoorn.

5.5 From Fishing Sector to (D1./D2.)

Fish and Shellfish // Algae

D1	Fish and shellfish	Fish Skin	Eco-leather
		Crabyon	Eco-fabric
		TômTex	Eco-leather/film
D2	Algae	Algae Sneaker Line	Eco-leather
		Algear	Eco-film/fabric
		Kelp	Eco-leather
		Carbo	Eco-leather
		Lichen	Eco-fabric
		Weaving Water	Eco-fabric

In the field of new eco-textiles made from fish and shellfish waste, we can highlight the group of eco-fabrics (Crabyon) and eco-leathers or films (Fish Skin, TômTex).

The seaweed sector can be a sector in itself, due to its multiple and varied biomaterial and bioenvironmental properties (CO_2 capture, energy production, colour variation), over and above its food consumption. We can also highlight the group of eco-tissues and/or eco-fabrics (Algear, Lichen, Weaving Water) and the group of eco-leathers (Algae Sneaker Line, Kelp, Carbo) (Fig. 5.4).

Fig. 5.4 Pixabay CC, by Fraugun

D1. | *From FISH to*

Fish Skin
Vivianne von Arx, The Netherlands, 2021
#fishskin #circulareconomy #sustainability #eggyoke #eco-leather
https://www.futurematerialsbank.com/material/fish-leather/

FRAMEWORK:
Designer Vivianne von Arx while working on this project reflected on transforming and bringing to life, new materials such as fish skin. For one ton of fish, we waste 40 kg of beautiful high-quality skin, a skin which can be transformed into leather. Made of only two natural ingredients, this material is wholly organic and biodegradable, sustainable and uses low energy to produce. The leather is waterproof and stable, the transformation from the skin to the leather includes sun and time. It opens doors to so many ways of using it while recycling the skin of a fish.

DESCRIPTION:
Creating fabric from fish skin through sewing involves a combination of traditional sewing skills and an understanding of the unique properties of fish leather. It is essential to prioritise quality and durability in the production process.

PRODUCTION:
Using a sharp object such as a shell, you clean the skin of the fish and remove the scales. The skin is then massaged with egg yolk and the organic soap helps the designer replace the water in the material with oil. The prepared skin is then dried in the sun and sewn to create large fabrics from the skin of different fish.

D1. | *From SHELLFISH to*

Crabyon
MAEKO and OMIKENSHI, Italy-Japan, 2019
#chitin #chitosan #circulareconomy #sustainability #biopolymer #eco-fabric
www.maekotessuti.com

FRAMEWORK:
Maeko combines the strong interest in the use of natural resources with the careful listening to the new needs of the fashion sector, especially in the field of ethics. It has the constant aim of enhancing the natural characteristics of the fibres, ennobling them through the use of modern finishing treatments, without ever distorting their naturalness.

In 2019 the historic worsted spinning company Filarte was acquired and became an integral part of Maeko in order to enhance and carry on from generation to generation a heritage of knowledge that is part of our country and of the manufacturing traditions of the Italian textile. Maeko wants to pass on ancient knowledge as an inheritance and in the custody of future generations, without renouncing innovation.

DESCRIPTION:
Crabyon is a fibre created by the Japanese company Omikenshi and recently used by Maeko, in addition to being antibacterial and antimicrobial, it is *haemostatic*, completely biodegradable, hypoallergenic, ecological and biocompatible. The antibacterial and antimicrobial functions of Crabyon are explained by the inhibition of the growth of bacteria and remain unchanged and permanent over time even after washing, use or other alterations by external agents. The Maeko company also uses many other natural fibres, in addition to the more common ones, such as pineapple, seaweed and lotus.

PRODUCTION:
The production process involves the crushing of crustacean shells from the food industry, mixing with bio-cellulose, without the use of solvents and subsequently the extrusion of this mixture. This method makes *Chitin* and *Chitosan* available, substances endowed with innumerable hygienic-sanitary properties, whose biocompatibility has been scientifically verified for use in the medical and pharmacological fields.

The fibre, in order to be prepared for spinning, requires a process that is obtained with the aid of combs. At the end of the preparation process, through the use of intersetting and the finisher, a bobbin is obtained which will then be mounted on spinning machines which will transform it into yarn. Finally, the yarn is wound into cones and will be used directly on the looms.

D1./A6. | *From SHELLFISH and COFFEE to*

Tômtex
Uyen Tran/New York, USA, 2020
#shellseafood #coffee #bioleather #circulareconomy #eco-leather #eco-film
www.tomtex.co

FRAMEWORK:

Due to the toxic process of tanning and dyeing leather, the Vietnamese joke that the trend colours of the coming season can be predicted through the colours of local rivers. Tanning hides and making synthetic leathers based on petroleum severely damages water resources and generates strong toxic elements: a study claims that workers in tanneries are much more likely to suffer from cancer. In coastal cities this industry dumps thousands of tons of waste into the sea. The food industry also produces strong environmental impacts by generating enormous amounts of food waste per year. TômTex was born with the desire to provide a sustainable alternative to harmful leather practices and create a safer work environment for employees through, precisely, the use of food-waste. Thus, TômTex contacted companies in Vietnam dedicated to collecting discarded shellfish remains to extract chitin, a fibrous polymer found in the shells and exoskeletons of marine fauna and which allows biological processes to be favoured in the production of biodegradable materials similar to leather for offer sustainable alternatives to the aforementioned harmful practices of the fur industry.

DESCRIPTION:

Tômtex is a plastic-free and 100% bio-based material, with high performances, created from shell seafood waste, coffee ground (and eventually with mushrooms waste), with the aim to work as a sustainable alternative instead of faux, synthetic and animal leather. It's 100% natural biodegradable and free of plastic, and it is often distinguished by its excellent softness to the touch while delivering high performance and durability. A coating of beeswax layer can enhance its water resistance properties. It can replicate any texture surface including animal skin/exotic skin textures and other design patterns.

PRODUCTION:

The Tômtex products use *chitin* or *chitosan*, a substance made, as said, from the shells of crustaceans combined with coffee grounds and/or with mushrooms waste to create a leather-like material. This "Chito Leather" can be used to create a variety of detailed and intricate textiles that have a high level of craft. By combining the chitin with the daily coffee grounds, a natural brown material is produced, 100% biodegradable. Unlike natural leather, which is permanently altered during the tanning process to prevent breakdown, the material generated can be naturally biodegraded or reintroduced into the production process to create a new material itself. The leather-like material is capable of reproducing exotic animal skin textures and can take on new and imaginative patterns, so it has vast potential. The project is currently in new research and development phases and seeks to collaborate with various brands to guide its future.

D2. | *From ALGAE to*

Algae Sneaker Line
NAT-2 ™ & Daniel Elkayam, Munich, Germany, 2011
#algae #vegan #circulareconomy #sustainability #eco-leather
www.nat-2.eu/collections

FRAMEWORK:
The nat-2™ project, as a brand and as a project, aims of creating a new awareness in the minds of consumers and the industry. The nat-2™ is all about innovation, design and sustainability paving the way for the future fashion footwear by using unique, uncommon materials such as milk, fish leather, natural felt, recycled leather and many vegan luxury alternative materials such as stone, wood, corn, cork, glass, fungus, coffee, grass, flowers, natural rubber.

DESCRIPTION:
Nat-2, is, in fact, an innovative high-end sustainable footwear brand bringing never used before natural materials into footwear design such as real stone, coffee grounds, hayfield, corn, cork, mushrooms, oxblood, fish leather, flowers, cannabis, red pepper, skeleton leaves, moss and many more. The Algae sneakers attempt to generate a perceptual change in the way that we act with our nature, by using bio sheets that made of dead algae picked from nature, and give them a new life cycle as a daily fashion item.

The sheets use different algae types that collected from nature and algae powders that cooked together and poured into moulds to dry up. The sheets are handmade and consist a variety of algae textures and colours that stimulate our senses.

The sheets can be reused by repeating of the cooking process and recasting it or to return it back to nature for biodegradation. The shoes are fully hand made in Italy, using regional and local vegan materials such as real cork insole, real rubber outsoles, vegan leather, glass and cruelty-free glue.

PRODUCTION:
The algae material was developed by Daniel Elkayam in Israel, while the sneakers were designed by Sebastian Thies in Germany and produced by hand in the nat-2™ workshop in Italy. The algae upper which makes up to 50% of the surface is accompanied by reflective glass and recycled pre-consumer PET bottles. The signature nat-2™ bio ceramic lining is German made and features real silver threads and ceramic fibres which are antibiotic, antistatic and work like an air condition. The insoles are made from real recycled cork and the outsoles are handmade in France by 99.9% natural milk rubber.

As a continuation of the SEAmpathy Project, the collaboration with nat-2™ is a way of expression of biophilic values, that seeks to create deeper connections and relations between man and nature. By these values a sneaker line of unisex 100% vegan and eco-friendly shoes with algae pattern.

D2. | *From ALGAE to*

Algear

Randa Kherba, Munich, London, 2019

#algae #bioleather #outdoor #sustainability #eco-film #eco-fabric

https://theface.com/style/randa-kherba-menswear-designer-arctic-man

FRAMEWORK:

The advent of synthetic materials such as polyurethane and Gore-tex has democratised access to extreme climates. Human interactions with these climates are temporary experiences, lasting days to weeks, but they have lasting repercussions for the landscape left behind. Waste is a complex issue on the mountains, where climbers frequently leave behind their gear to avoid carrying any extra weight for a safer descent. There is a need to design gear specifically for short-term use: made from nature, to be experienced in nature.

DESCRIPTION:

Purely by existing, we consume energy in the natural environment. The point of Algear is to highlight a move away from the mainstream notion of brutal energy extraction and move towards a mindful form of consumption. The ability to cultivate microalgae allows for a sustainable form of extraction.

This project harnesses the organic growth assembly of cyanobacteria, a blue-green alga which is among the oldest known phototropic organisms. Their photosynthetic abilities and adaptiveness to stressful conditions offer potential for a material that can endure the shifting/changing climates. Their simple cell structure allows the species to grow at a fast rate, making them almost infinitely renewable.

PRODUCTION:

The project imagines a future where outdoor gear can protect us from the elements and be created by elements that already exist, or have been left over from the past. Tents, jackets and sleeping gear would be made from algae. This project contributes to the importance of the leave-no-trace code, through growing a textile that naturally biodegrades and nourishes the land in the process, allowing for the conservation of an ecological future through our transient adventures outdoors.

In theory, customers could come to Kherba with an order and, depending on how long they would need the product for, the designer would estimate the growth period and customise the product for how much usage they'd get from it. And to top it all off, the product could be left out in the wilderness to biodegrade.

D2. | From ALGAE to

Kelp

Jing-cai Liu, The Netherland, 2020
#seaweed #bioleather #outdoor #sustainability #eco-leather
http://jingcailiu.com/seaweed-matters/

FRAMEWORK:
In this project, seaweed is explored to be a possible future material that could replace current materials that are not sustainable such as plastic and leather. Seaweed is an organic material that is of vegetable origin, and it will decay quicker than plastic or animal leather.

DESCRIPTION:
In the first phase, a variety of material explorations were explored to make bioplastics from seaweeds. After explorations, a material of seaweed was created that is comparable to leather.

In the last phase, workshops and interviews with seaweed experts were conducted to collect insights into this topic. Additionally, a video was created for provocation and to raise debates about using seaweed material in the future.

PRODUCTION:
The seaweed leather material is made by soaking the seaweed in materials that are used for making bioplastics such as vinegar, glycerine, etc. After combining different materials in the soaking processes, a good piece of seaweed leather is created that is soft, strong and smooth. In the second phase, three designs are made with this seaweed material. These are possible products that could emerge in a nature-based future. The products include the following wearables: A pair of shoes, a bag and a cardholder. The idea is that this seaweed leather could replace animal leather in the future. In some way, it could be used now for making products that require less sturdy materials. Although the final version of the seaweed leather is made by only using diluted glycerine, the steps of using different components are more significant for creating a good piece of seaweed leather that is soft, strong and smooth.

D2. | *From ALGAE to*

Carbo
David Cabra, Colombia, 2021
#charcoal #eco-leather #biodegradable #newmaterial #sustainability #circulare-conomy
https://davidcabra.com/carbo

FRAMEWORK:
CARBO presents as an innovative bio-derived skin, showcasing a sustainable and biodegradable solution distinct from traditional leather and synthetic materials originating from petroleum. Its unique composition combines activated charcoal with sodium alginate, derived from Sargassum, an algae species proliferating along the coastlines of Latin America. Utilising Sargassum not only employs a renewable biomass but also promotes the principles of a circular economy. Enhanced with activated charcoal, CARBO boasts superior characteristics including odour control, pollutant absorption and UV protection, marking its significance in the fashion industry.

DESCRIPTION:
The conventional production of leather and petroleum-based textiles significantly contributes to environmental degradation within the fashion industry. Leather processing consumes vast quantities of water and energy, with its tanning procedures emitting hazardous chemicals like chromium into the surroundings. Equally, synthetic fibres, such as polyester, necessitate the extraction of finite resources, releasing greenhouse gases and toxic pollutants during manufacturing. In contrast, CARBO's production is characterised by minimal energy consumption and the absence of noxious elements, rendering it an eco-conscious alternative. Moreover, incorporating Sargassum in its production aids in managing its excess, which otherwise inflicts extensive ecological and economic damages on the ecosystems of Latin America. The formulation of CARBO also offers a potential pathway for carbon capture, effectively sequestering carbon dioxide when integrated into bioplastic polymers. Distinguished by its tactile and functional attributes this bio-skin differs from other bioplastics that contain activated carbon.

PRODUCTION:
The procedure initiates with the cleansing, drying and milling of Sargassum seaweed into a refined powder. This powder is subsequently combined with water and heated to dissolve the contained sodium alginate. Following filtration to eliminate contaminants, sodium alginate is precipitated through the addition of calcium chloride. This powder is then amalgamated with water, forming a viscous, gel-like substance. To enhance its malleability and durability, glycerine and soybean oil are infused into this mixture. Finishing the composition, activated charcoal powder is thoroughly mixed in. This final bioplastic blend is then distributed into moulds, mirroring the intended design and dimensions. The moulds undergo a drying process at gentle temperatures to remove moisture and allow the bioplastic sheets to harden. Upon drying, these sheets are extracted from the moulds and trimmed to their specified sizes.

D2. | *From ALGAE to*

Lichen
Mirte Luijmes, Sweden, 2022
#lichen #biofabric #circulareconomy #sustainability #eco-fabric
https://www.futurematerialsbank.com/material/lichen-wool-cotton/

FRAMEWORK:
In this material-driven research within the field of textile design, the starting point of this work was the realm of lichens. Lichens are packed and fragile in their dry state: when they come into contact with water they soak up the water like a sponge, become greener, expand and become flexible. The lichens adjust and react to the circumstances of their surroundings: humid or dry. Luijmes' project is driven by the lichens and adjusted by their properties in the knit which make them the bio-informer. The knitting technique, the chosen yarns and the colours are considered to achieve the optimal interaction between the two elements informed by the properties of the lichens. Furthermore, the knit is designed in a way so the lichen can enter and be part of the textile. Design decisions are guided by the natural process and this makes nature an active collaborator in the work which makes the lichens co-worker to the knitting technique and the designer.

DESCRIPTION:
A lichen is a symbiosis between algae and fungus. A lichen survives because of this symbiotic relationship and together the fungus and the algae create the property of adjusting to their environment. Lichens are important to our ecosystem because of their diversity, they indicate air quality, they serve as food for big animals or shelter for insects and they protect substrates against acids. The responsive property of the lichens to humidity led to the investigation of this species in combination with the flexible properties of knitted textiles. The knit enhances the lichens' properties in its stretchable, textural and colour possibilities.

PRODUCTION:
Working towards a material library, the transformable properties of the textiles are being explored. Lichen in combination with knit is investigated in four categories: transformation, shape, texture and dye. The lichen and the knit adapt their role according to the category and demonstrate the potential of their various collabo-rations in each category. The lichens combined with natural fibres such as wool, paper yarn and cotton are biodegradable and can be given back to nature to increase the lichen population. These fibres absorb water which makes the lichens have more moisture and gives them the chance to photosynthesise and grow. Due to the slow growth of lichens, the material is not introduced as a design material but is used to raise awareness by visualising the existence of the lichens. In addition to arousing appreciation for the unseen species, the work aims to let the viewer reflect upon the relationship of humanity towards nature.

5.6 From Mushrooms and Bacteria to (E1./E2.)

Mycelium // Bacteria

E1	Mycelium	Ephea	Eco-leather
		Fungi Narratives	Eco-fabric
		Mylo	Eco-leather
		The Pure Hyphae Project	Eco-leather/film
		MuSkin	Eco-leather
		We Grew Together	Eco-leather
		Eco Warrior	Eco-leather
E2	Bacteria	reGrow	Eco-leather
		Kombucha Couture	Eco-leather
		Scoby-compo	Eco-leather
		A Baby, A Beast	Eco-leather
		Moving Pigments	Eco-dyes
		Maqui Biotextile	Eco-leather

In the field of new eco-textiles made from fungi and bacteria, we can highlight the group of mycelia eco-fabrics (Fungi Narratives), but especially the group of performative eco-leathers and/or light films (Ephea, Mylo, The Pure Hyphae Project, MuSkin, We Grew Together, Eco Warrior).

The field of bacteria is another important sector, due to its multiple and varied bio-environmental properties (good maintenance, ability to evolve, etc.). We can highlight the group of eco-leathers (reGrow, Kombucha Couture, Scoby-compo, A Baby-A Beast, Maqui Biotextile) and the group of eco-dyes produced using bacterial properties (Moving Pigments) (Fig. 5.5).

Fig. 5.5 Pixabay CC, by khfalk

E1. | From MYCELIUM to

Ephea
SQIM/Mogu Srl, Italy—2022
#mycelium #fermentation #biofabrication #circulareconomy #sustainability #eco-leather
https://www.sqim.bio/, https://www.ephea.bio/

FRAMEWORK:
SQIM is a company partnering with natural agents (i.e. fungi) to ferment a renewed and responsible material culture, as driven by cutting-edge natural technologies rooted in regeneration. By growing selected species of fungal mycelium, the intricate network of filamentous fungal cells SQIM literally cultivates innovatively biofabricated materials and products, contributing to harmonising human-driven activities with the rhythms and functioning of the larger ecosystem. This, by embracing the inherently regenerative skill of fungal mycelium, which, fed through organic residues and by-products from other value-chains, is guided through tailored fermentation-based processes to engineer and structure a variety of materials suitable for diverse applicative purposes.

DESCRIPTION:
Engineered to achieve a low ecological footprint with minimum energy inputs needed, ephea is grown from and consisting of fungal mycelium, the vegetative body of mushrooms. Thanks to a proprietary fermentation technology and to a deep understanding of the underlying bio-technological processes, mycelium is therefore grown as raw material in bespoke static bioreactors, and then stabilised through a dedicated downstream process, allowing for the creation of high-quality products embedding a fundamental shift from an extractive/exploitative approach to a thoroughly (re-)generative production paradigm rooted in bio-fabrication.

PRODUCTION:
Ephea demonstrates that innovative natural materials hold the factual promise for the creation of a near future where human activities and the rhythms of the larger ecosystem are not in conflict with each other. Ranging from an appreciation for its premium haptics, to its overall consistency along production, ephea has received multiple recognitions from highly demanding industrial players exposed to the product, evaluating it as the leading mycelium-based alternative. This is for instance the case when looking at Balenciaga's Maxi Hooded Wrap Coat realised with ephea, presented in March 2022 at Paris Fashion Week and later commercially launched in October 2022 in selected Balenciaga's stores globally. Altogether, with ephea, SQIM has set the grounds to contribute to the creation of a novel fashion industry responsibly rooted in regenerative processes, as driven by the desire to literally engage with life and for life and contributing to strongly mitigating environmental impacts while advancing sustainable practices, without compromising on performance, quality and aesthetics.

Credits: ephea™ high-quality raw material: 100% mycelium/fungal biomass—©SQIM/ephea

E1. | *From MYCELIUM to*

Fungi Narratives

A. Fedder, S. Fiorentino, M. Hansen and M. V. Jørgensen, Denmark, 2019
#fungi #biodegradable #circulareconomy #sustainability #eco-fabric
www.competition.adesignaward.com

FRAMEWORK:

A walk in nature inspired the work with fungi and was the kickstart of the Fungi Narratives research project. This research shows a possibility to extract Fomes Fomentarius from nature without negatively impacting the ecosystem, as well as the potential of growing that species artificially. The team conducted over 100 experiments studying the fungi and combining them with bioplastics. The duality between the man-made and the natural was the main motivation driving the experiments, which can have a positive impact on the fields of sustainable product and fashion design.

DESCRIPTION:

Fungi Narratives is an experimental material research project. The results are extracted fungi materials and biodegradable bio-polymers used in implementations for fashion design techniques for garments, accessory design and expanding to print and communication design with paper production. Most interesting was the use of the Amadou inside the fungus, which could be used as an alternative to leather, with limited negative environmental impact. Amadou can have an impact in the future of sustainable alternatives to leather.

The project included harvesting the fungi, treating and dissecting them with the purpose of using abundant resources from nature to create materials with a possible positive sustainable impact on the future of design. The project started in November 2019 and ended in September 2020.

All team members are graduates from the Design School Kolding's MA Program "Design for Planet" with backgrounds in communication and fashion design. The project was started as part of a material research course in the program and then developed further by the team in September 2020. It was then exhibited at the Dutch Design Week 2020 in Eindhoven as part of the exhibition "Materialised".

PRODUCTION:

For the up-cycled fashion pieces we used techniques such as knitting, beading and creating a filament from the natural resources. Furthermore, editorial design of the presentation book, letter pressing and hand binding was done also implementing the material library developed. The Amadou face mask is 20 cm in length and 11 cm in width, made from the inner section "Amadou" of Fomes Fomentarius Fungus. The side straps that hold the mask up were made of wool.

E1. | From MYCELIUM to

Mylo

Bolt Threads, San Francisco—Portland—Arnhem, USA—The Netherlands, 2020
#fungi #biodegradable #circulareconomy #sustainability #eco-leather
https://boltthreads.com/technology/mylo/

FRAMEWORK:
Mylo is a sustainable leather alternative made from mycelium, the root-like system of mushrooms. As a material, Mylo delivers the timeless aesthetic and luxurious feel of leather, but without the planetary impact associated with raising livestock. While cows require extensive resources and years to raise, the mycelium used to make Mylo is grown in less than two weeks inside a state-of-the-art vertical farming facility powered by 100% renewable electricity.

DESCRIPTION:
Mycelium—the root-like system of fungi—has evolved for billions of years beneath the forest floor. To make Mylo, the company brought this infinitely renewable resource to the surface—engineering a unique process to grow and transform mycelium into a sustainable alternative to traditional and fully-synthetic leathers.

PRODUCTION:
The process begins with mycelium cells grown on a bed of renewable, organic substrate inside our vertical farming facility that is powered by 100% renewable electricity. Billions of cells, or hyphae, form a densely interconnected network of soft foam that is harvested to make Mylo. The remaining substrate is composted. Guided by Green Chemistry principles, the harvested mycelium is processed and transformed into sheets of soft biomaterial. Surface texture and finishing is applied by a Leather Working Group (LWG) gold rated tannery for an unmistakable resemblance to traditional leather.

E1. | *From MYCELIUM to*

The Pure Hyphae Project
Annah-Ololade Sangosanya, FabLab BCN, Spain, 2022
#mycelium #biodegradable #circulareconomy #sustainability #eco-leathe #eco-film
https://www.annahololade.com/the-purhyphae-project

FRAMEWORK:
Considering that various micro and macro-organism, such as fungi and more specifically their mycelium, are capable of biodegrading the main components of textile (cellulose and more complex plastic molecules) an opportunity to rethink the linearity of the textile industry emerges. Beyond breaking down waste products, the mycelium hyphae network can produce mycelium-based materials, including leather-like materials, adoptable in the fashion industry. This project investigates ways to produce flexible mycelium materials through the biodegradation of various combinations of denim textile waste, synthetic textile waste, food waste and spent coffee grounds. The mycelium used was from the Pleurotus ostreatus (oyster) fungi. The results show that P. ostreatus (oyster) mycelium grows on all the combinations of food waste (vegetable peels and coffee grounds) with textile waste (synthetic textile and denim textile) and even grows on denim textile waste only. However, the mycelium did not entirely degrade the fibres but only partially digested it, leading to a leather-like composite made of the mycelium and the remainder of its substrate. Provided the soft nature of the substrate, the textile waste and food waste mycelium composite is also malleable, and therefore interesting for further textile applications.

DESCRIPTION:
A protocol for post-processing of the flexible composite material using low energy and natural components (heat, water, glycerol and beeswax) was created to make a composite leather-like fungal material. The whole process of partial biodegradation of textile and food waste mixes, followed by post-processing, is thus a sustainable process. It enables a circular way of treating textiles, therefore, closing the loop of the current linear model, offering an opportunity to get rid of poorly recycled waste and reducing the associated environmental impacts.

PRODUCTION:
Textile waste is shredded and mixed with grounded coffee waste, humidified, put in a petri dish and sterilised in an autoclave. Once sterilised, the mix is inoculated with mycelium and put to grow in an incubator at 30 °C, high humidity. Once the mycelium has colonised the entire substrate, it can be post-processed into a composite material. Composite is heat pressed, plasticised in a glycerol bath and coated with wax.

E1. | From MYCELIUM to

MuSkin

Pangaia Grado Zero S.r.l., Firenze, Italy, 2017

#mycellium #circulareconomy #sustainability #eco-leather

https://lifematerials.eu/en/shop/muskin

FRAMEWORK:

The industrial processes associated with the manufacture and treatment of animal skins to produce clothing or leather present a very important environmental impact given the large amount of toxic chemical products used and the waste produced, thrown on land or into rivers and seas. In this context, it is essential to propose sustainable alternatives to traditional methods.

DESCRIPTION:

Developed by the Italian company Grado Zero Espace (specialised in the study of innovative materials and technologies for the development of new manufactured products) and then carrier forward by Pangaia Grado Zero, a spin-off company, MuSkin is a 100% vegetable peel alternative to animal leather. It comes from the *Phellinus Ellipsoideus*, a kind of big and large parasitic fungus that grows in the wild and attacks the trees located in the subtropical forests. MuSkin is a soft material with a suede-like touch however, its consistency and texture, may vary from soft to slightly harder than cork.

PRODUCTION:

The total absence of toxic substances and chemical products in its preparation makes MuSkin ideal for close-to-skin applications and, being totally natural and rich in endemic penicillins, it could limit the proliferation of bacteria. MuSkin has the capacity to absorb moisture and then to release it in a short time, just like a fabric. It is not waterproof in its natural form, but it can be treated with ecological wax. In fact, this product is highly breathable and can be waterproofed by treating it with ecological waxes.

The material may not have a smooth surface due to its entirely natural origin. Each piece is a unique organic product. It is strongly recommended to laminate or coupling MuSkin with other backing materials such as fabrics or papers, because this increases its mechanical resistance and helps, for example, with the stitching or gluing in the production articles. The absence of toxic substances in the material and its ability to absorb moisture and release it in a short time—just like fabric—make the product suitable for use in applications such as shoes, gloves or watch straps. The dimensions of each piece are unique and variable, and the production is adapted to small series productions.

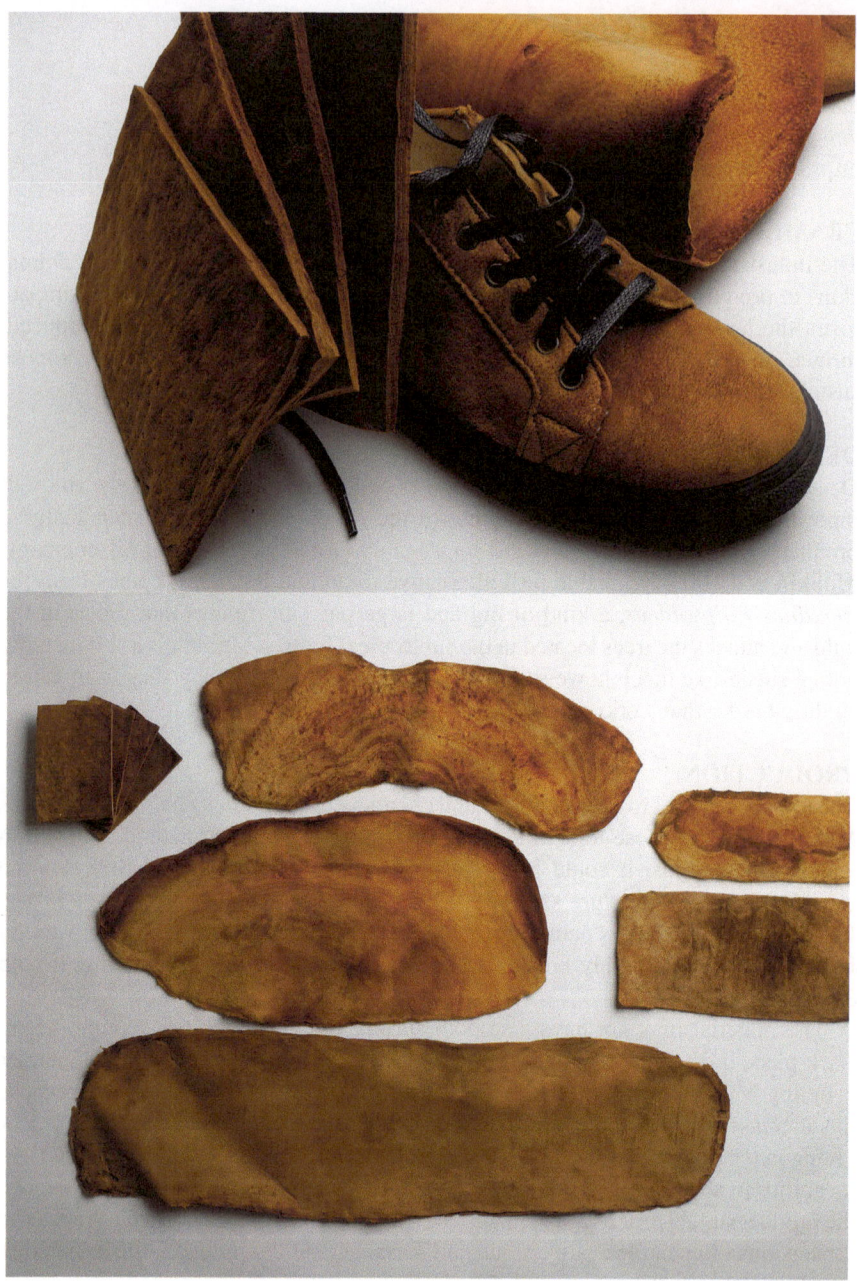

E1. | *From MYCELIUM to*

We Grew Together
Mari Koppanen, Romania, Norway, 2019
#fungi #biodegradable #circulareconomy #sustainability #eco-leather
https://marikoppanen.com/amadou

FRAMEWORK:
We Grew Together is a culturally sensitive exploration of amadou, a suede-like material derived from Fomes fomentarius, also known as tinder mushroom. This material is traditionally used for crafting hats and bags in Transylvania and is deeply rooted in local folklore. Amadou has been used by various cultures throughout Europe and Asia for a wide range of purposes, such as starting fires, wound dressing, and as a spiritual substance, and it possesses absorbent, anti-septic and isolative properties, and is also cruelty-free.

DESCRIPTION:
Tinder fungus was once a valuable resource that was not only gathered from the forest but also cultivated with great success. During the 1800s, the major manufacturers of tinder fungus were in German cities such as, Frankfurt, Strasbourg and Ulm. Later on the use of the material dwindled and knowledge of it began to fade. Nowadays, Transylvania is perhaps the only region where this material tradition still persists. We Grew Together aimed to discover new ways to use amadou while also preserving the disappearing tradition of amadou handicrafts. Drawing inspiration from the cultural patterns of the village, the handcraft and the historical significance of the material, the designer created a 3-piece collection that highlighted the unique qualities of amadou. The resulting objects fused the natural characteristics of amadou with the region's rich cultural heritage, using the same techniques as the local artisans.

PRODUCTION:
Amadou preparation is entirely done by hand, starting with harvesting mushrooms during the summer and late autumn. To create amadou, a soft and flexible layer found inside the mushroom is peeled, trimmed and stretched. This delicate layer is carefully separated from the cuticle and pore tubes and stretched using circular motions before being left to dry. Although the preparation process may appear straightforward, it demands extensive practice and knowledge. Choosing the appropriate mushrooms is crucial and recognising the areas where to pick them is important. Even though you may find a group of 20 mushrooms growing on the same tree, it's best to only pick a few, and the harvesters also respect the forest and the trees. The skilled amadou artisans master specialised techniques to enhance the processing, ensuring that they obtain the largest possible sheets of amadou.

E2. | *From BACTERIA to*

reGROW

Surzhana Radnaeva, Benjamin Denjean, France—Spain, 2018
#kombucha #bacteria #circulareconomy #sustainability #teamushroom #eco-leather
https://traditionalfutures.org/clothes-design#/kombucha/

FRAMEWORK:

"My fascination with materials has always been profound. Four years ago, I stumbled upon the concept of Leather made from Kombucha, a revelation that struck a nostalgic chord within me. As a child growing up in Siberia, I vividly remember the Чайный Гриб—the 'Tea Mushroom'—which I consumed regularly. Fast forward to the present, and I found myself experimenting with small pieces of Kombucha leather. However, it wasn't until the Reshape competition that the breakthrough occurred. Thanks to Vivien Roussel's cultivation, I obtained a large 2 × 2 m piece, enabling me to design two garments for the reGROW project. The journey from childhood memories to the European Kombucha drink, from Kombucha leather to garments, and now to potential Medical Kombucha gauze, epitomizes the transformative power of 'Reshaping'".

DESCRIPTION:

Design is an agent for values and vision. reGROW is an exploration in sustainability, local-manufacturing and the intelligent use of a living organism—kombucha. Kombucha, a bacterial cellulose, is one of the most promising natural, renewable polymers for biomedical applications. reGROW utilises the simple and rapid cultivation of kombucha, common maker tools, reuse of waste grape crops and open source philosophies to deliver an intelligent biomedical gown that can assist in the rehabilitation of patients with a selection of skin injuries and ailments.

The reGROW project involves a multidisciplinary, international team in the investigation of bacterial cellulose potential as an assistive and reactive fabric with use-specific applications. The water retention and moisturising properties make the material an ideal candidate to combine both burn treatment and the delivery of encapsulated active compounds onto damaged skin. reGROW is a smart, yet fashionable, scalable and easy-to-produce medical gown.

PRODUCTION:

reGROW opens up two main tracks of future development: finely-tuned, personalised drug delivery and on-demand, easily producible skin treating gauze for remote areas and provides an option in biomedical treatments for local communities around the globe.

E2. | *From BACTERIA to*

Kombucha Couture
Sacha Laurin, Davis, California, USA, 2015
#veganleather #sustainability #kombucha #bacteria #eco-leather
http://www.kombuchacouture.com

FRAMEWORK:

Kombucha Couture is a live culture vegan leather design company based in Davis, California, and Kombucha Jewellery and Couture are unique creations from Sacha Laurin whose career as a professional cheesemaker has taken her from fermenting milk with bacteria and yeasts to fermenting green tea with SCOBYs (Symbiotic Colony of Yeast and Bacteria) and transforming the growing "colony" into live clothing and fashion.

Sacha's goal is creating and furthering this new versatile fabric that can mimic leather, canvas, silk or butterfly wings depending on growing and production techniques. Her love for colour and the magical play between light and texture has inspired her to use food dyes which keep the live culture breathing and vibrant.

DESCRIPTION:

Kombucha Couture is creating biodegradable vegetal non-violent leather and probiotic textile that is unique, cutting edge and good for you from the inside out.

PRODUCTION:

Sacha Laurin, owner and founder is combining her passion for and knowledge of fermentation and dairy science, and her love and vision for slow and organic fashion. She grows sheets of cellulose-based kombucha textile from tubs of fermented green tea, the bacterial and yeast SCOBYs spinning cellulose around themselves as silkworms spin silk. The kombucha cellulose is then dried into vegan leather that still contains the probiotic, detoxifying organisms and can imitate canvas, leather or silk, depending on the process used. Laurin combines kombucha textiles with repurposed metals and materials and other environmentally friendly textiles such as bamboo fibre. The result is a new, sustainable resource of huge potential and application.

E2. | *From BACTERIA to*

SCOBY-compo
Riina Õun, United Kingdom, 2020
#kombucha #sustainability #bacteria #circulareconomy #eco-leather
https://www.riinao.com/crafting-organic-waste-for-fashion

FRAMEWORK:
As an alternative to currently available PVC- and PU-based "vegan leather" products, designer Riina Õun has developed a new material—SCOBY-compo. It utilises the waste bacterial cellulose generated by local kombucha drink producers, reprocessing these cultures into a more sustainable, vegan and leather-like material. For ethical reasons, people often opt for non-animal-derived products made of so-called vegan leather, unaware that many vegan leather products are made from PVC- and PU-based plastic which are very harmful to the environment.

DESCRIPTION:
Through rigorous and meticulous experimentation and the use of natural oils, waxes and organic compounds, the developed material is water-resistant, flexible and strong. The smell is enhanced by using essential oils to develop a pleasant and unique scent. The result is a fully commercial, market-ready product that can be created in large quantities as a viable alternative for the fashion industry.

PRODUCTION:
To evidence the viability of SCOBY-compo, Riina Õun crafted a collection of handbags and purses that each demonstrate a different technique for production, from traditional stitching to modular assembly and liquid moulding. Working closely with local industries, the aim is to create a fully circular, closed-loop system where the organic waste material can be harvested, processed, sold and eventually home-composted at the end of its life cycle, and rather than contaminating the environment, will nurture it.

E2. | *From BACTERIA to*

A Baby, A Beast

Mari Koppanen, Oslo, Norway, 2022
#bacteria #sustainability #kombucha #circulareconomy#eco-leather
https://isola.design/Designer-Projects-A-Baby-A-Beast

FRAMEWORK:

Designer Mari Koppanen has used SCOBY in various ways in her artistic Ph.D. research project as a substitute for plastic or leather and plays with different scales. A Baby, A Beast investigates the line between the human and non-human through microbial cellulose. The study presents a range of SCOBY material samples in different shapes, sizes, and both living and dried formats. The research strongly emphasises the symbiosis between human and microbial world: our bodies are home to a diverse community of microorganisms.

DESCRIPTION:

SCOBY has a human skin-like texture, which evokes a sense of caring for it by providing warmth, nutrition and oxygen—the necessary elements it needs to grow. Koppanen proposes microbial cellulose as a material that is not made to last forever, but rather to decompose. She demonstrates it as an alternative to materials and objects that are difficult to recycle and often designed for short-term use, such as plastic sequins.

PRODUCTION:

Fermentation is started by combining water, sugar, tea and a piece of SCOBY. The type of tea used, such as black tea or green tea, will affect the colour of the final material. Sugar serves as a food source for the microorganisms in the kombucha. To avoid contamination, it is important to use clean and sterilised equipment, using filtered water, and storing the kombucha in a clean place. The material is allowed to grow for a period of 7–21 days, depending on the desired thickness. Then it is harvested and washed with soap. Dyeing is best done directly after harvesting by placing the material into a dye batch for 1–2 weeks. The material dries for 3–4 days. During the drying process, it will lose approximately 80–90% of its thickness. Once it is dry, it can be cut into the desired shape and size. During the kombucha fermentation process, a cellulose-based biofilm forms on the liquid. This floating microbial mat, SCOBY, contains mainly cellulose, different bacterial species and various types of yeasts. During fermentation, the bacteria and yeast consume sugars and other nutrients in the liquid and produce acetic acid, alcohol and carbon dioxide as by-products. The SCOBY is weaved on the air surface as a result of the accumulation of these microorganisms. The exact composition of SCOBY will depend on the specific microorganisms present in the culture and the conditions of the fermentation process. It is 100% compostable and even edible. All samples presented are dyed with natural pigments from plants, lichen and fungi.

E2. | From BACTERIA to

Moving Pigments
Charlotte Werth, United Kingdom, 2022
#bacteria #sustainability #newmaterial #circulareconomy #eco-dyes
https://charlottewerth.com

FRAMEWORK:
Moving Pigments aims to scale up and automate the process of co-designing textile patterns with pigment-producing bacteria. Bacteria dyeing is a rather beautiful and unique method of dyeing, creating colour gradients and lines when guided, which cannot be imitated easily. Nevertheless, the microbes grow in slightly unexpected ways and are part of the design process. It intends to enlarge and make visible a reality that is usually hidden from sight, showing us the incredible beauty of this parallel microscopic world.

The high degree of uniformity demanded in the context of mass production and consumer capitalism has led to extensive usage of petrochemical dyes. These often have disastrous impacts on ecosystems through the pollution of water courses and landscapes. In contrast, bacteria dye has many environmentally friendly advantages, including far lower water usage and no use of harmful chemicals. Placing this method within the industry's context is desirable and necessary to provide an alternative to the destructive status quo.

DESCRIPTION:
By centring living organisms as an integral part of a collaborative production process, the outcome can be explicitly designed but never foreseen precisely. Co-designing and co-producing with microorganisms means understanding their way of growing and applying that when generating patterns. The aim of this work is to explore the possibility of reproducing this predictably unpredictable practice on a larger scale.

PRODUCTION:
Challenging the established separation of human and non-human species can create meaningful innovation. Designing with and not against nature requires alternative practices and new instruments. The machine developed within this work is designed to experiment and explore the process of bacteria dyeing through automation. It represents a case study for the prospective large-scale implementation of sustainable co-designing dye practices.

Credits: Dying Machine, by Tom Mannion

E2./F1. | *From BACTERIA and PLANTS to*

Maqui Biotextile
LABVA, Chile, 2019
#maquitree #sustainability #biotextile #circulareconomy #eco-leather
https://makeitcircular.whatdesigncando.com/projects/maqui-biotextile-by-labva/

FRAMEWORK:

Laboratorio de Biomateriales de Valdivia (LABVA) recognises the potential of our territory associating ethnobotanical ancestral knowledge to bio-fabrication processes. The Maqui tree is native to Chile and it is extremely abundant as it is a pioneer species, this means that it's the first to colonise degraded soils, setting the perfect conditions for other native plants to grow. It is also a food source for birds that depend on this tree to survive their migratory routes. The Maqui also plays a key role in the Mapuche tradition and cosmovision; it is used as food, medicine and natural dyes. Creating a material 100% derived from this tree microbiota is both a discovery and also valorisation of all the scales involved in its creation.

DESCRIPTION:

This native biomaterial materialises the collaboration in its ecosystem by colonising and feeding other beings, the collaboration between communities through openly sharing their ancestral knowledge of its conscious and sustainable collection, and the collaboration of its symbiotic microbiota, since it's the bacteria and yeast present in its leaves and fruits that do this microscopic weaving. LABVA ensures that it grows, feeding this culture and creating the perfect environment for it to thrive. This results in a biomaterial that is deeply rooted in its territory; a biomaterial that is conceived, makes sense and can only be grown in the south of Chile. They seek to create a heterogeneous, diverse and local biomaterial palette to promote sovereignty and territorial autonomy. Promoting values through the creation of emotionally binding materials in order to change our throwaway culture. 897.

PRODUCTION:

Wild fermentation of leaves and fruits of *Aristotelia chilensis* tree. The process can take up to 3 months to create the first culture.

- Microbiota extraction: Harvesting fruits and leaves to ferment.
- Fermentation: It takes up to 3 months to develop the first batch of wild bacterial cellulose. The bacteria and yeast present in the fruits and leaves initiate a process in which yeast breaks down sugars and bacteria start spinning nanofibres of cellulose.
- Harvesting and processing: The cellulose is harvested and processed with bio additives and natural pigments (if needed) that will stabilise the material for an even desiccation process.
- Finishing: The fabric is then coated with natural oils and waxes. Rigorous quality control measures would be implemented to ensure that the fabric meets industry, standards for strength, colourfastness and overall quality.

5.7   From Plants to (F1.)

Plants

F1	Plants	Eco Warrior	Eco-leather
		InterWoven	Eco-fabric
		Maqui Biotextile	Eco-leather
		Lovr	Eco-leather
		Climafibre	Eco-fabric
		Flocus	Eco-fabric
		Latex	Eco-leather
		Bio-Invasive Textile	Eco-fabric
		Juhla	Eco-fabric
		Mader	Eco-leather
		Fique	Eco-fabric
		Simbiosis	Eco-fabric
		Yerma	Eco-leather
		Desserto	Eco-leather

In the field of new eco-textiles made from different plant waste, we can first of all highlight the group of eco-fabrics (InterWoven, Climafibre, Flocus, Bio-Invasive Textile, Juhla, Fique, Simbiosis), but especially the group of performative eco-leathers (Eco Warrior, Maqui Biotextile, Lovr, Latex, Mader, Yerma or Desserto).

Proposals capable of introducing the multiple capacities of some specimens of plants used for food, infusions, nutritional supplements, etc. (Fig. 5.6).

Fig. 5.6 Pixabay CC, by ignartonosbg

F1./E1. | *From PLANTS and MYCELIUM to*

Eco Warrior
Studio Silvio Tinello, Argentina, 2016
#mycelium #yerbamate #biomaterials #circulareconomy #eco-leather
www.materialdriven.com/blog/2018/6/20/a-union-of-grown-materials-by-silvio-tin
ello

FRAMEWORK:
Biology is today the new technology that allows designers to be a new type of
material interpreters: the interaction and creation between organic dialogues and
unforeseen matters represent the starting point for exploring new alternatives. Interesting processes take place in these crosses, resulting new evolved, biodegradable
and bio-made, materials.

DESCRIPTION:
The Silvio Tinello Studio "harvests" lighting fixtures, clothing and accessories. He
developed a material from the waste of yerba mate—a type of holly plant which
is used as a national drink in Argentina—to create sustainable fashion garments. It
is a South American tree whose leaves, when processed, serve as a tea drink. It is
traditionally drunk in countries such as Argentina, Uruguay and part of Brazil. An
important part of the work carried out focuses on the combination between yerba
mate and fungal mycelium. The idea of designing using biology as technology makes
it possible to "cultivate" and "grow" products, instead of manufacturing them, thus
promoting a paradigm shift in the design universe. Objects designed in this way
represent an opportunity to make a statement of intent and contribute to a more
sustainable design based on biology.

PRODUCTION:
The bio-manufacturing process starts from a fungal bio-agglomerate composed of
the discarded yerba mate and bound with mycelium, the vegetative part of the fungi.
On the other hand, bacterial cellulose results from the cultivation of bacteria that
transform sugar into cellulose. This cellulose makes it possible to generate flexible,
translucent pieces of various dimensions that can be stamped, engraved, moulded,
etc.
    The combination of two materials makes possible to take advantage of the proper-
ties of each one of them, strategically combined in various formulas and applications.
The designed objects have a combination of hard and soft components and both,
the bio-agglomerate and the cellulose, are moulded, cast, sewn and bent to meet the
appropriate needs, in each case. The fungal bio-agglomerate used (or organic *telgopor*
or *styrofoam* "in Creole") is a logical evolution of the rubber and resin-based materials developed in the first research projects generated, but with the advantage that the
new material is biodegradable and returns to the environment once its life cycle has
ended. There is also a strong local, regional or national cultural bond with the natural
materials and textures used that adds value to the new bio-manufactured objects.

F1. | From PLANTS to

InterWoven
Diana Scherer, The Netherlands, 2016
#plantroots #sustainability #biotextile #circulareconomy #eco-fabric
https://dianascherer.nl

FRAMEWORK:

In the project, InterWoven, the natural network of the root system turns into a textile-like material. Using templates as moulds, the roots are channelled, forming the new material. During the growth process, the roots conform to the patterns and the root material weaves or braids itself. The material is biologically manufactured and degradable.

DESCRIPTION:

It is created from subterranean living biomass, which continues to produce itself as long as there are enough nutrients, light and warmth. CO_2 storage in living biomass is seen as one of the ways to reduce CO_2 concentration. Plants absorb CO_2 and fix the carbon in organic compounds.

PRODUCTION:

- Selection of Plants with Root Fibres: Certain plants may have fibres in their roots that are suitable for textile production. Examples might include plants with long, strong and flexible roots.
- Root Extraction: The roots of selected plants are harvested and processed to extract the fibres. The extraction process may involve cleaning, separating and treating the fibres to make them suitable for weaving.
- Spinning and Weaving: The extracted root fibres are spun into yarns, and these yarns are then woven into fabric. The weaving process can vary, with options for different patterns and textures in the final carpet fabric.
- Dyeing and Finishing: If colour is desired, the fabric may undergo a dyeing process. Finishing processes can be applied to enhance qualities such as softness, durability and resistance to wear.
- Quality Control: The produced carpet fabric undergoes testing for various properties, including strength, colourfastness and overall quality. Quality control measures are implemented to ensure the fabric meets industry standards.
- Manufacturing Carpets: The root-based fabric is then used in the manufacturing of carpets. This involves cutting, shaping and binding the fabric into the desired carpet design.

F1. | *From PLANTS to*

Lovr
LOVR, Geleen, Germany, 2021
#hempresidues #sustainability #vegan #circulareconomy #eco-leather
https://www.revoltech.com/#lovr

FRAMEWORK:
The production of leather is extremely harmful to the environment. In addition to a high carbon footprint in the process and the use of harmful chemicals in its development, leather also carries the stigma of animal suffering. The greatest difficulty for the textile industry is to develop a truly sustainable alternative to this material. Unfortunately, greenwashing is a common practice in the field. Most vegan leather options on the market today have plastic and other petroleum-based substances in their composition. Furthermore, the mixture of materials makes the products non-recyclable.

DESCRIPTION:
LOVR is produced from waste from German hemp cultivation. It's 100% biodegradable, vegan, petroleum and chemical-free, and the first purely plant-based product with a leather like look, feel and durability. Compared to real leather it saves 99.7% of CO_2. Customers can specify the colour and surface embossing and process the product with standard sewing machines. LOVR will be distributed across Europe to the fashion, furniture and automotive sectors. The start-up is supported by an EXIST grant at TU Darmstadt, where they are leading the development of the material until it is ready for the market It is the true leather alternative that consists of 100% plant material and has a minimal ecological footprint. The textile looks and feels like leather. The single-layer structure allows for dyeing and a variety of colours. LOVR is also malleable and non-abrasive. The unique composition technology allows the product to be fully recycled. The textile is also biodegradable and is free of chemicals and plastic. LOVR's manufacturing process requires only 0.3% of the CO_2 emissions used in leather production.

PRODUCTION:
The production of LOVR is based on a regional circular economy. LOVR's manufacturing is carried out from residues of hemp cultivated in Germany. The plant is primarily grown for medicinal and nutritional purposes. LOVR takes care of the by-products. Other advantages of hemp are that it has a high density of cultivation, it is an efficient carbon sink, and its cultivation is increasing significantly in Europe.

LOVR is sustainable and the technology behind its manufacturing is scalable. The result is a single-layer material that requires neither a substrate nor a coating and therefore offers customers many benefits in terms of variability and processability.

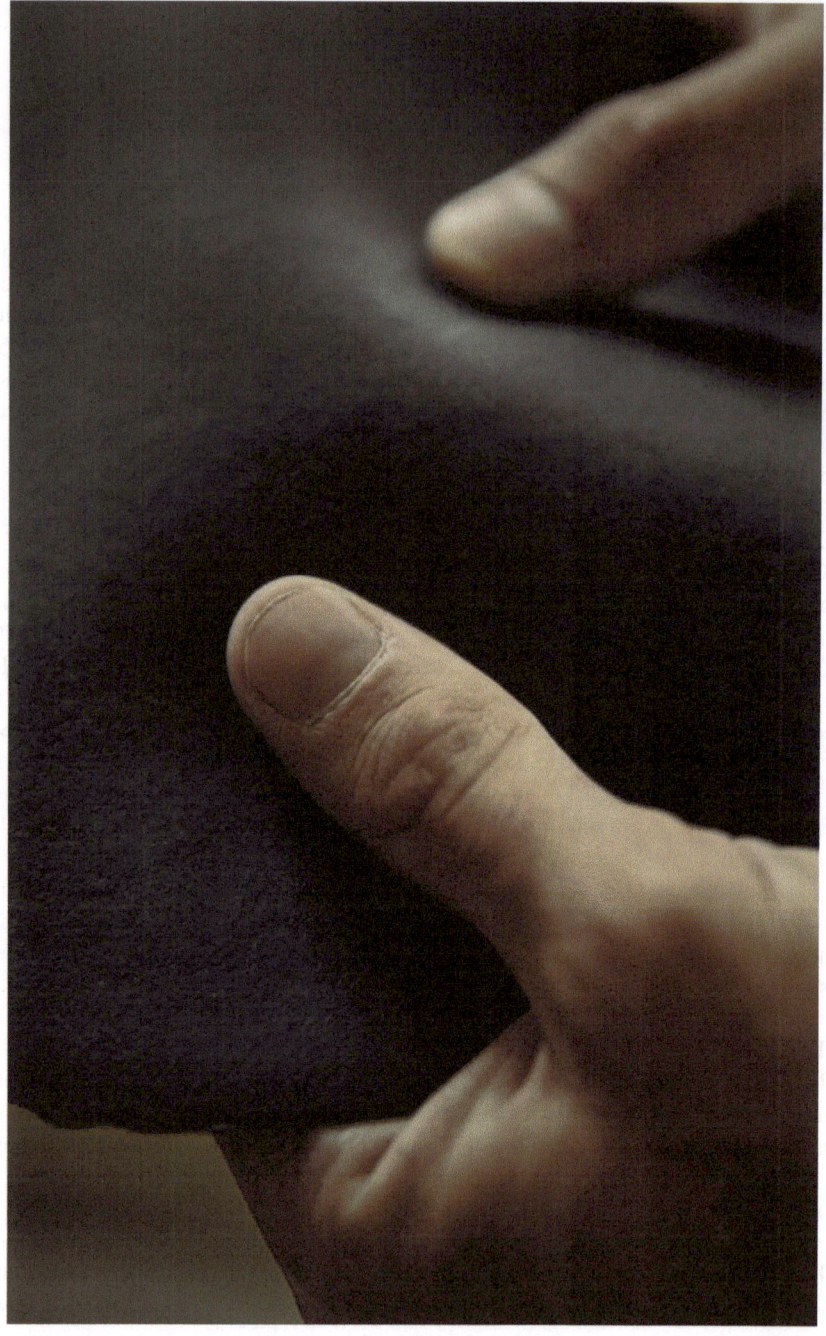

F1. | *From PLANTS to*

Climafibre
Jessica Redgrave, United Kingdom, 2023
#sunflower #sustainability #biotextile #circulareconomy #eco-fabric
https://jessredgrave.com

FRAMEWORK:
Climafibre uses sunflowers to develop a range of modular solutions for the fashion industry that supports regenerative food systems, protects biodiversity and aids climate mitigation. The fashion industry's environmental impact is disastrous and consumerism has fuelled the desire for fast fashion, which is reliant on the consumption of finite resources and intensive farming practices. Increasing amounts of fertilisers and pesticides are needed to meet these demands, degrading the soil, which inhibits regeneration, resulting in a loss of arable land. Jess Redgrave found a solution to this through sunflowers use, which are utilised as a part of regenerative agricultural systems and aid climate mitigation through soil remediation and boosting biodiversity above and below ground, hey can be grown without fertilisers and can be companioned and rotated with other food crops. Sunflowers have extensive root systems that allows symbiotic relationships with beneficial bacteria, fungi and microbes to be formed, promoting healthy soils. Sunflowers can withstand drought and can grow in vastly varying ecosystems. Their natural resilience has made them a model for scientists studying climate change adaptation.

DESCRIPTION:
Climafibre has developed fibre for textiles, natural dyes and a hydropic coating made entirely from sunflowers. Climafibre uses sunflowers to develop a range of modular solutions for the fashion industry that supports regenerative food systems, protects biodiversity and aids climate mitigation.

PRODUCTION:
Climafibre has developed fibre for textiles, natural dyes and a hydropic coating made entirely from sunflowers.sing enzymes derived from bacteria and fungi, Climafibre has worked closely with scientists to develop a unique process to isolate cellulose fibres from sunflower stems. These fibres are then combed and spun into a yarn, and then woven into a fabric. The hydrophobic coating is made from a by-product of the sunflower oil industry and provides water-resistant protection for natural fibres without the use of harmful chemicals. This coating allows the fabric to maintain its breathable qualities with minimal alteration to its aesthetics or hand feel. Climafibre's bold colour palette has been developed from pigments extracted from various parts of the flower.

F1. | *From PLANTS to*

Flocus

FLOCUS™, The Netherlands—China (supporting office)—Italy, 2021
#kapok #sustainability #vegan #circulareconomy #eco-fabric
https://www.flocus.pro

FRAMEWORK:

FLOCUS™ presents a range of kapok textile materials, which provide the textile industry with a naturally sustainable and regenerative alternative that has not been available before.

Kapok Fibre is a pure and natural, non-food, vegan product. Increasing the demand of this material, in the countries where kapok trees grow, helps to involve communities and develop economic prosperity—as well as having a positive impact. It's worth mentioning that Flocus™ operates the sole factory that has mechanised the extraction of fibre from kapok pods, setting them apart as pioneers in integrating social responsibility into the kapok industry. It's crucial to explain that kapok is not an artificially created fibre. The fibre is already in the fibre form in its pods which grow as fruits, therefore it is not a timber product. Kapok fibre grows on regenerative trees, on non-agricultural land.

Flocus™ offers a new environment involvement by creating the demand for fibres and therefore for more trees to be planted and halt deforestation, in line with the SDGs of the UN. By planting kapok trees with KRAF (Kapok Regenerative Agriculture Forestry), FLOCUS™ generates an efficient eco-system which avoids erosion, avoids deforestation, sequesters carbon and increases O2 in the atmosphere, preserves water, supports poly-cropping, organically fertilises the land.

DESCRIPTION:

Kapok fibre is 100% biodegradable and recyclable although it would need more energy to recycle than to utilise new fibre. In every application, during the production process, and at the end of the product's life, kapok is regenerative. Indeed, over-consumption of kapok actually helps the planet by increasing the number of kapok trees and reducing deforestation. FLOCUS™ is suitable for use in fashion, performance fabrics, for use in workwear, home textiles, automotive, medical and other industrial products.

PRODUCTION:

FLOCUS™ is leading the industry to a better way for the people and the planet. It is a pioneer in the development of a textile fibre, kapok, whose production is environmentally friendly, and whose end-use products naturally boast the best in versatility and functionality. FLOCUS™ fibre may be spun into yarn, to add value to any blend. Kapok can be blended with organic cotton, Better Cotton Initiative (BCI) cotton, lined, recycled polyesters and other materials. Flocus™ is continually innovating and exploring additional blends in order for the resulting yarn to benefit from the comfort and performance features associated with kapok. Its performance also excels in nonwovens (or paddings) as Kapok provides the same warmth as down, naturally substituting any down jacket padding with the same or better functionality.

F1. | From PLANTS to

Latex

Magdalena Sophie Orland, The Netherlands—Italy, 2020
#latex #sustainability #biotextile #circulareconomy #eco-leather
https://www.magdalena-orland.de

FRAMEWORK:

The MaDe workshop and competition series—based on the master project BETWEEN_SPACES—served the deepening and further study of contemporary lace by using sustainable materials for the manufacturing process. Lace is a material with filled and empty spaces. Therefore, the background of the material always plays an important role. Certain properties of lace were investigated with natural latex on an experimental level and already-gained knowledge was used to take the project to a new standard.

DESCRIPTION:

The material natural latex is extracted from rubber plants. With a few additives, it becomes an extremely elastic and resilient material. Natural latex is pourable and extrudable, which makes it particularly suitable for the production of contemporary lace. For the textile and fashion industry, its strong elasticity is a positive aspect. In combination with other natural materials natural latex offers a wide range of experiments that allow the typical appearance of latex goods to be skipped and visually redefined. Natural latex is a purely biological product. It is CO_2 neutral, biodegradable and free of harmful substances. Thus the circularity can be guaranteed.

PRODUCTION:

Multi-stage pouring processes enable differentiated colouring of flat yet openwork textiles. It dries translucently and can be dyed well. The short drying time allows quick processing. In interaction with light, exciting surfaces and a variety of patterns are created. Natural latex can be excellently extruded, which enables a new approach.

In combination with copper wire, the materials change parts of their properties and can be shaped for a long-term effect. By using conductive wire, externally controllable movement and heat mechanisms can be developed.

The main countries of origin are Malaysia, Indonesia and Thailand, which is a disadvantage from a transport point of view. European spurge plants could be investigated in the next step and similar properties could be generated.

F1. | From PLANTS to

Bio-Invasive Textile Library
Xue Chen, United Kingdom, 2023
#neddle #sustainability #naturalfabric #circulareconomy#eco-fabric
https://www.gp-award.com/en/produkte/bio-invasive

FRAMEWORK:
The Bio-Invasive Textile Library is a win–win solution for both the ecology and fashion worlds. The project uses LISI (London's) invasive plants as raw materials for fibres and dyes to physically prevent the loss of biodiversity in the region and the inhumane treatment of animals. Its bio-invasive furs are explored to achieve zero waste dyeing and technical production processes. By exploiting the textiles wasted during dyeing and technology development, further develop related ancillary bio-invasive materials.

DESCRIPTION:
The study conducted a comprehensive inventory of invasive plants in London and comparing it to the local resources and their distribution map was created. The project exclusively employed invasive plants from the region for all dyeing procedures, thereby making a significant contribution to invasive plant management and the mitigation of biodiversity loss in the area.

PRODUCTION:
In the technical part, new textiles are created by innovating diverse technologies such as "implanting", "felting" and "spinning", providing fashion designers with solutions in the form of a "library" of the Bio-invasive Textile Library as a real fashion solution for fashion designers. Regarding the dyeing process, a distinctive approach was adopted in contrast to the conventional practice of incorporating chemical components. Instead, clubmoss was employed as a substitute for chemical mordants, and various invasive plants were utilised for multiple dyeing sessions, resulting in a diverse range of colours. Additionally, the fibre component of the project incorporated invasive plant-based materials such as ramie fibre and nettle fibre. The craftsmanship aspect involved the utilisation of innovative machines, namely the "Implanting" machine and the "Flocking" machine. The Implanting machine represents a groundbreaking technology capable of integrating any type of fibre into any fabric. The needle mechanism, simulating crochet patterns, enables the precise placement of fibres at the back of the fabric. Notably, this portable machine permits the implementation of various patterns and designs, facilitating versatility in fibre implantation. The Flocking machine, on the other hand specifically tailored to accommodate the unique characteristics of nettle and ramie fibres and allows the transformation of nettle and ramie fibres into felts akin to woollen coat fabrics. By combining these two machine technologies with knitting and spinning techniques, as well as employing a regenerative production process, the project successfully realises the amalgamation of diverse fabric forms.

F1. | From PLANTS to

Juhla
Lena Ringel, Germany, 2022
#pineneedle #sustainability #newtxtile #circulareconomy #eco-fabric
https://lenaringel.com/juhla

FRAMEWORK:
70% of all trees in the forests of Brandenburg State, Germany are pine trees. Due to the sandy soil and low annual precipitation, pines have a much easier time thriving here than other tree species. Pine wood is one of the most important timbers in Germany and Europe. Millions of cubic meters are felled in Germany every year for a variety of uses as construction material. It is also a popular wood in furniture construction due to its decorative structure. But pine trees are not only made of their wood: One of their best-known features is the evergreen pine needles. These are produced as waste products during logging. That's estimated millions to billions of pine needles per tree that go to waste, a billion pine needles filled with potential.

DESCRIPTION:
With the help of warp yarn, the pine needles can be woven together to form a completely biodegradable textile: When it is damaged or no longer wanted, it can be composted and provide the basis for new pine trees to grow.

PRODUCTION:
Named after the Finnish word for feast and celebration, "Juhla" represents a new and sustainable way to celebrate Christmas. The rug—made from 100% domestic pine needles—replaces the traditional Christmas tree in a durable, sustainable and compact way. The falling needles are collected, sorted and reconnected. The process itself is a small celebration of the cycle of life.

Turning pine needles into fabric involves several steps. Here's a general overview of the process. Harvesting of mature pine needles that are fully grown and have fallen naturally. Cleaning the pine needles by removing any dirt, insects, or other impurities gently brushing them. Sun-drying: Allow the cleaned needles to air-dry in the sun to ensure they are completely free of moisture. Softening: Submerge the dried needles in water for a period of time to make them more pliable. This process helps in softening the needles and makes them easier to work with. If the needles are too dry they may break during further processing.

Weaving: Each needle is individually placed during the weaving process. This can be done using traditional weaving techniques or by incorporating the pine needles into existing fabric structures.

F1. | From PLANTS to

Mader
David Cabra, Colombia, 2021
#pinesawdust #sustainability #bioskin #nowaste #eco-leather
https://davidcabra.com/mader

FRAMEWORK:
MADER, a bioskin material, represents an innovative leap forward in the quest for environmentally friendly alternatives to traditional leather and petroleum-based textiles. Created from pine sawdust, a by-product of industrial wood processes, and sodium alginate derived from Sargassum seaweed, this material addresses significant environmental concerns by making effective use of renewable resources. By leveraging pine sawdust and sodium alginate, MADER not only reduces industrial wood waste but also contributes to the management of seaweed overgrowth, which poses ecological challenges along Latin American coasts the synergy of these materials offers a promising application in sustainable material development, emphasising environmental benefits and resource efficiency.

DESCRIPTION:
The industrial application of pine sawdust, especially its lignocellulose content, has proven advantageous in boosting this bioskin material's mechanical properties. The utilisation of pine sawdust in MADER is especially significant due to the beneficial properties of lignocellulose, which enhances the material's tensile strength and durability. Lignocellulose, a complex biopolymer, has been shown to improve performance metrics such as tensile strength when incorporated into bioplastics. Additionally, the biodegradability of wood sawdust is advantageous, as it decomposes over time, in contrast to petroleum-based materials that can persist in the environment for centuries. Sodium alginate, derived from Sargassum seaweed, aids in coastal environmental management by limiting the excessive growth of this seaweed, thus benefiting marine ecosystems.

PRODUCTION:
By offering a sustainable, vegan and biodegradable alternative to leather and petroleum-based fabrics, it meets the increasing consumer demand for environmentally responsible products while setting a new standard for material ecology. This bioplastic formulation involves blending sodium alginate derived from Sargassum with water, glycerine and natural oils to produce a homogeneous slurry. This base mixture is then combined with wood sawdust to enhance its mechanical properties. The final blend is cast into moulds and cured at room temperature, resulting in a robust bioplastic.

In summary, MADER is a cutting-edge development that embodies both environmental responsibility and economic practicality. Its evolution from forest and marine waste materials not only addresses critical industrial and environmental concerns but also paves the way for future advances in sustainable materials.

F1. | From PLANTS to

Fique
Rosana Escobar, Colombia, 2021
#fiquefiber #wastereutilisation #circulareconomy #sustainability #eco-fabric
www.dezeen.com/2021/12/15/rosana-escobar-unravelling-coffee-bag/

FRAMEWORK:
Fique, a versatile and sustainable natural fibre derived from the agave plant, has become a focal point in the production of eco-friendly textile products. This resilient material, native to the Andean region of South America, particularly Colombia and Ecuador, is known for its exceptional strength, durability and environmentally friendly characteristics. It provides the fibre used to produce coffee bags exported from Colombia to the rest of the world. The high demand has resulted in a large fique cultivation and industry which also promotes sustainable agriculture practices, as the plant requires minimal water and grow.

DESCRIPTION:
Using the material in its raw form before its industrially turned into a yarn, showed the unique qualities of the material. The fluff, a by-product of the fibre, was explored for their robust and durable quality, making them well-suited for a variety of applications like tudy ropes to durable bags, or carpets, rugs and tapestries when threated with felting techniques to create a skin-like material. Fique products are known for their resistance to wear and tear, ensuring a long lifespan and reducing the need for frequent replacements. The production of fique fibres often involves local communities and traditional craftsmanship, fostering social and economic development.

PRODUCTION:
Fique fibres are extracted from the leaves of the agave plant through a labour-intensive process, emphasising the artisanal nature of the production. Once the fique leaves are harvested, the next step involves extracting the fibres. This is traditionally done manually by skilled artisans. The outer skin of the fique leaves is removed to reveal the long, tough fibres within. The extracted fibres undergo a process called retting, which is a method of separating and softening the fibres from the other plant components by soaking the fibres in water or exposing them to moisture, allowing the natural processes of decay to break down non-fibre components. After retting, the fibres are dried to prevent mould and ensure that the fibres are ready for subsequent manufacturing stages. The dried fique fibres are spun into yarn or thread. Traditional spinning methods involve the use of spinning wheels or spindles, although modern manufacturing may incorporate mechanised spinning processes. If colour is desired, the fique fabric may undergo a dyeing process using natural or synthetic dyes. The fabric is then finished, which may involve treatments to enhance softness, durability, or other desired characteristics. The finished fique fabric is used in the manufacturing of various end products, such as bags, carpets, fashion accessories and home decor items.

F1. | From PLANTS to

Simbiosis
Alejandra Ortiz de Zevallos, Lima, Peru, 2020
#commonreed #sustainability #biotextile #circulareconomy #eco-fabric
https://www.futurematerialsbank.com/material/common-reed/

FRAMEWORK:

The common reed (phragmites australis) is a plant that can reach 4 m in height, and the stem can be up to 2 cm in diameter. It can be found in humid soils, on the banks of rivers and lagoons. They can live in both brackish and freshwater. Many consider that the reed is a pest due to the speed with which it proliferates anchoring the rhizome/stolons in the ground. From that moment on, the underground rhizomes begin to grow, this extends horizontally through the territory and, in turn, the stem of the reed grows vertically, on which leaves grow. Finally, the leaves die and fall to the ground and only the stems remain, the canes and these take between 3 and 4 years to fall and then become part of the weakened one.

DESCRIPTION:

In Lima the reed population is scarce, since the city has grown, burying its rivers and water channels. Today there are only two operational canals, Surco and Huatica, both usually very contaminated. The reed grows there in invisible spaces, in the midst of urban overcrowding, on the margins of the invisible hydraulic fabric. Weaving the leaves from the common reed is an exercise of trying to extend life—an act of nourishing and being nourished by this fibre. To construct these living sculptures, Alejandra combines two weaving techniques, crochet and q'eswa. Crochet is an originally Western practice, while the q'eswa is a type of high Andean braiding used since pre-Hispanic times to construct bridges, the only one left alive is the Qeswachaka bridge. A technique she learned from the community of Kacllaraccay in Cusco, and then has applied that knowledge to the common reed, which worked well because of the strong resilience that characterises this plant.

PRODUCTION:

The process to work with the reed material is long and laborious. First, you have to harvest the leaves and investigate the riverbanks to identify any sector that has a high reed population. After collecting the leaves, they must be washed, since mainly the rivers that cross the city are normally very polluted and it is important to disinfect the leaves before working with them. After washing them, you have to dry them in the sun for 3 or 4 days, then these leaves are divided into several strips so that you can later build a rope with them. The q'eswa or rope is a braid with two strands, each strand turns on itself and with the other, to facilitate turns and twisting it is preferable to have wet hands and maintain a constant rhythm. When all the collected reed leaves have already been transformed into a long rope, then you begin to crochet the sculptures.

F1. | From PLANTS to

Yerma
Bionuma Lab—UNTREF—Ana Laura Cantera/Buenos Aires, Argentina, 2017
#yerbamate #fruitwastematerials #circulareconomy #sustainability #eco-leather
https://www.analauracantera.com.ar/yerma

FRAMEWORK:

Bionuma Lab is a biomaterials laboratory founded by Ana Laura Cantera at Universidad Nacional de Tres de Febrero, based in Buenos Aires, which experiments with materials and creates design objects using biological techniques, sustainable processes and organic waste such as yerba mate, human hair, fruit waste and mycelium.

Bionuma's orientation (a line of biomaterials, but also a platform and a research laboratory) combines a scientific approach and a creative vocation. The proposed approach wants to be innovative and disruptive and leans towards productive aspects associated with nature and the territory through the combination of digitisation, electronics and natural organisms.

DESCRIPTION:

Founded by Ana Laura Cantera and based in Buenos Aires, Bionuma produces new bio-composite materials and also objects made using biological techniques, sustainable processes and everyday organic waste.

Since June 2017, Cantera has been working with Emiliano Gentile (founders of Mycocrea Lab) on researching, training and creating objects from new materialities, particularly with mycelium (living fungal hyphae), kombucha and yerba mate bioplastic.

The laboratory works investigating new solutions under the triad design-biology-sustainability from the principles of the circular economy.

PRODUCTION:

One of the products emerged from the laboratory is Yerma: a 100% biodegradable biomaterial similar to leather, made from the residue of used yerba mate. Flexible and resistant, it can be both elastic and rigid. It can be sewn, drilled, laser-engraved and works as a non-stick coating for 3D printer bases. From the ground yerba mate itself, the team has generated functional sandals that have proven to be an excellent sustainable material for making footwear (exhibited in 2019 at the MIT-3.0 Global Community Bio Summit).

Credits Ph. @leogphotographer

F1. | *From PLANTS to*

Desserto
Adriano di Marti Sa., Guadalajara, Mexico, 2019
#cactus #fruitleather #circulareconomy #sustainability #eco-leather
www.desserto.com.mx

FRAMEWORK:
Leather market is a multibillion business throughout the world and many people are concerned about its environmental (and ethical) impact. The reports from the United Nations Environment Program (UNEP) point out that fashion is one of the most damaging and polluting industries for the environment with textile production responsible for 20% of global wastewater and 10% of global carbon emissions. In this context it is important to investigate innovative processes to find sustainable alternatives to traditional animal leather.

DESCRIPTION:
In 2019 Adrián López Velarde together with Marte Cázarez founded the company Adriano di Marti Sa. in Mexico with the aim of developing the first cactus-based material in the world as an alternative to leather, complying with rigorous environmental and quality standards. Desserto is a plant-based vegan leather made from nopal (cactus) proteins and fibres: it is a cruelty-free and a polyvinyl chloride (PVC)-free leather alternative, distinguished by its great softness to the touch while offering great performance for a wide variety of applications. Cactus vegan leather is partially biodegradable and has the technical specifications required by the fashion, leather goods, luxury packaging and furniture industries. Nopal does not involve animal sacrifices and not need irrigation water; it is a very efficient natural carbon sink and does not pollute waterways like the conventional leather industry.

PRODUCTION:
The mature nopal cactus leaves are manually harvested every 6–8 months. The cactus is unharmed in the process and the same plant can keep producing for 8 years. The leaves are then cleaned and ground into a pulp which is dried under the sun for 3 days. Through an industrial process proteins and fibres are extracted from the pulp and transformed into a bio-resin, which is used to make the synthetic leather biomaterial. Desserto's process which uses no toxic chemicals and 90 percent plant-based materials. The material can be applied in a variety of products including handbags, footwear and car interiors through collaborations with companies such as H&M, Adidas and Mercedes Benz. Other accessories are created with Desserto by Mexican artisans and designers: they include clothing such as jeans, coloured shirts and also accessories such as necklaces, sunglasses and many more. These items go very well with the different seasons and occasions, thanks to the textures, geometric shapes and colours they have.

5.8 From Packaging and Other to (G1./G5.)

Plastic // Gum // Paper and Carton // Hair // Various

G1	Plastic	Rifò	Eco-fabric
		Plastex	Eco-fabric
		Plastigela	Eco-leather
G2	Gum	Gomma	Eco-fabric
		Gum-Tec	Eco-leather
G3	Paper and Carton	Consumption of Heritage	Eco-fabric
		Abitinuovi	Eco-fabric
G4	Hair	Human Material Loop	Eco-fabric
		Contemporary Hairwork	Eco-fabric
		Weaving Water	Eco-fabric
		Wolfwall	Eco-fabric
G5	Various	Made In	Eco-fabric

In the field of new eco-textiles derived from various types of food packaging waste and other components of packaging waste (Plastic // Gum // Paper and Carton // Hair // Various), we can first mention the group of eco-fabrics (Rifò, Plastex, Gomma) or eco-leathers (Plastigela, Gum-Tec) obtained using Pack-Plastics or Gum.

Recycled packaging papers and cartons or various mixed components allow to conceive eco-fabrics (Consumption of Heritage, Abitinuovi, Made In).

Another unique organic material that can be processed is hair, which is difficult to use for food because of the bacterial effects of keratin, but which is particularly suitable—as an organic compostable element—to create biodegradable eco-fabrics (Human Material Loop, Contemporary Hairwork, Weaving Water, Wolfwall) (Fig. 5.7).

Fig. 5.7 Pixabay CC, by torstensimon

G1. | *From PLASTIC to*

Rifò
RIFÒ, Firenze, Italy, 2018
#wastetextile #plasticpackaging #circulareconomy #recycle #eco-fabric
www.rifo-lab.com

FRAMEWORK:
Every year around 2 million t-shirts are sold around the world. It is, perhaps, the most purchased garment of all time. Most of these t-shirts are made of cotton. Cotton is an inexpensive but "thirsty" fibre: for a simple cotton t-shirt, an enormous amount of water is required, something like more than 2,000 L.

Within the fast fashion industry, t-shirts are often worn a limited number of times and then thrown away after a short period of time from purchase.

Another important problem today is the enormous use of plastic bottles, which most of the time are not recycled correctly. Around ten million of these containers end up floating in our oceans, disrupting the planet and marine life. Putting these two bad habits together, you can see how both lead to dramatic consequences for society and the environment. In these circumstances the question is whether it is possible to produce a sustainable T-shirt. A 100% recycled T-shirt can be a solution.

DESCRIPTION:
The RIFÒ textile firm has created a T-shirt—Rifò Upcicled—that wants to be sustainable, responsible and social: a durable garment that makes the wearer feel good. Using an innovative system based in the regeneration of cotton and plastic bottles collected from the sea. Each t-shirt is made from 1 kg of cotton scraps and 4–5 plastic bottles.

As has been pointed out, virgin cotton is one of the most polluting fibres in the textile industry. Many liters of water and many pesticides are used to produce a single T-shirt.

Instead, with this method, the RIFÒ T-shirts require only 30 L of water and help to the recycling processes.

PRODUCTION:
The recycled shirt provides benefits and comfort as well as great advantages. *RIFÒ Upcycled*, is an entirely recycled, sustainable and circular product.

A recycled t-shirt is a simple t-shirt made from 100% recycled cotton.

From the cotton waste and the collected and crushed plastic bottles, a new fibre is created, with which the tissue and fabric of the garment is created.

Although somewhat more expensive, the shirt has enormous added values,

Buying a cheaper garment to be used rarely and ending up in the trash in a short time means making decisions of low environmental quality, often associated with deficient working conditions in less developed countries.

G1. | *From PLASTIC to*

Plastex
REFORM STUDIO, Cairo, Egypt, 2011
#plasticbags #packaging #circulareconomy #recycle #eco-fabric
www.eformstudio.net

FRAMEWORK:
One of the main current objectives today is to design with a responsible, socially and environmentally positive ambition. *Reformstudio* is a brand and a lifestyle at the same time. Immediately after the 2011 revolution, many creators and entrepreneurs from the new Egypt were particularly involved with the great change that was intuited. The feeling of responsibility to serve society and the environment provided great motivation to solve important common problems. In this sense and as is known, the negative impact of plastic bags on the environment is enormous.

In Egypt—and many other countries—it is the second most wasted material. Most of this waste ends up in landfills, in the sea or simply burned, generating toxic elements.

These dynamics associated with excessively standardised and competitive industrial production require new, more involved solutions with a better collective habitat.

DESCRIPTION:
Plastex is a new designed material made from reused plastic bags. The idea is to prolong the life cycle of plastic bags before it gets labelled as "trash". The average period of usage of a single-use plastic bag is only 12 min. By looking at plastic bags as a raw material rather than waste, Reform Studio has been able to transform it into a new durable eco-friendly handmade fabric. The new material is designed to raise awareness about waste and the possibilities behind reusing what was once destined to become "trash".

PRODUCTION:
Plastex project proposes a new ecological fabric that transforms discarded plastic bags into a new 100% handmade ecological tissue generated from the reused plastic bags themselves and integrated with cotton or, eventually, polyester threads, prolonging the life cycle. Socially, the company contributes to reviving the weaving or textile industry in Egypt, by promoting crafts and empowering local communities (especially disadvantaged women with limited resources and education).

Through Plastex they are taught new skills, incentivising possible more sustainable income to help families. Production takes place in the company's workshop after collecting, sorting and turning the plastic bags into threads for later weaving on handlooms. The possibilities and potentials are limitless for the beautiful and efficient material generated that continues to be developed through hundreds of new patterns that are applied in a wide range of applications such as furniture, home accessories and fashion accessories.

G1. | *From PLASTIC to*

Plastigela
Soowon Chae, the Netherlands, 2020
#plastic #ochre #circulareconomy #recycle #eco-leather
https://www.soowonchae.com/plastigela

FRAMEWORK:
Plastigela is a new material consisting of recycled plastic, ochre, gelatine, glycerine and water. Designer Soowon Chae has used recycled plastics as the main ingredient to create a new kind of textile material which has distinguishable texture and colour. Plastic is one of the most critical global issues to be solved and it is threatening the ecosystem and marine life. How to reuse and recycle plastics became one of the biggest challenges to designers.

DESCRIPTION:
Normally recycled plastics are utilised by designers through the method of melting and moulding into a new form. However, he used the recycled plastics as artificial pigments by grinding them and making them into tiny small particles. Each side of Plastigela has different colours and textures. One side has a bright colour and rough texture because of the plastic particles and the other has a natural earthy colour and smooth texture because of the ochre.

PRODUCTION:
This material can be hand-stitched or machine-sewed, so it can be used as a normal textile material. And he is continuing the research exploring how to apply the material to the products and textiles.

G2. | *From GUM to*

Gomma

Bianca Streich, Germany, 2022
#chewinggum #sustainability #polymers #circulareconomy #eco-fabric
https://www.futurematerialsbank.com/material/chewing-gum/

FRAMEWORK:
GOMMA is a material exploration that critically questions consumer behaviour and material flows in the contemporary food industry and urban environment. Components of chewing gum include gum base, sweeteners, plasticisers, flavourings and colourants. The gum base may contain up to 46 different chemicals, including synthetic polymers derived from crude oil such as paraffin and petroleum waxes as well as styrene-butadiene, vinyl acetate and polyethylene polymers to maximise elasticity. These components are non-nutritive, non-digestible, water-insoluble and foremost non-degradable. More often than not chewing gum is being dumped to pollute urban environments by the thousands. Even after long periods of time exposed to the elements, the chewing gum can effectively be restored to its former elastic state.

DESCRIPTION:
Discarded chewing gums made of non-degradable, petroleum-based synthetic polymers litter urban areas. Their existence has become an integrated feature of the cityscape worldwide, frequently found on the seats of public transport, sprinkling the pathways or sticking to monuments. Some countries have started to fine these illicit public disposals or have altogether banned the selling of chewing gum. The chewing gum, is the material base used for GOMMA, the gum was recovered, amongst other sources, from the historic landmark of the Berlin Wall at Potsdamer Platz in Berlin, which is now greatly covered in chewing gum. This phenomenon started several years ago, distorting the memorial value of the site.

PRODUCTION:
GOMMA is exploring the potential that this material has, due to the properties of the elastomers that make up the chewing gum base. Once the matter is heated in a warm water bath, it can be cleaned and easily reshaped. Through this process, the material loses its age-related impurities and discolourations and even recovers its fresh menthol smell. At this point in the process, colourants can be added. Manual kneading and threading techniques are then used to produce a smooth, malleable dough which can be transformed into weavable strings. Letting the material air dry in intervals reduces its stickiness and makes it easier to handle. The dried gum strings can subsequently be woven or crocheted. Besides the methods of textile manufacturing, other processes such as cutting, moulding and casting are also applicable. As previously mentioned, the material's properties allow it to be recycled several times by relooping the heating, cleaning and shaping processes.

G2. | From GUM to

GUM-TEC
Gumdrop Ltd., London, United Kingdom, 2009
#chewinggum #circulareconomy #sustainability #injectionmoulding #eco-leather
https://gumdropltd.com/gumtec/

FRAMEWORK:
350 billions of chewing Gum are consumed in the world every year, 30 million in Italy. Huge figures, which have a direct impact on the environment and on sustainability, since in many cases chewing gum ends up on the street, where collecting it is a long and expensive undertaking. Cleaning just fifty centimetres of asphalt takes at least half an hour and costs between 50 cents and 2 euros. To avoid this, Anna Bullus, a London designer, has created a new business around chewing gum and its recycling.

DESCRIPTION:
GUM-TEC is born from the will of Anna Bullus to find a way to transform chewing gum into objects of daily use. To do this, she took advantage of her university studies by discovering that the main ingredient of this product is rubber base, commonly known as synthetic rubber, a type of polymer similar to plastic. And as such it can be used in the same way. So, she founded Gumdrop LTD to collect used chewing gum and reborn it in a new sustainable material, GUM-TEC®.

PRODUCTION:
The creation of the rounded, bubble-shaped pink bins invites the "chewers" to throw the gum inside them. The containers themselves are produced from GUM-TEC. Above each bin it is explained that the Gumdrop Bins and collected gum is recycled to produce new Gumdrop Bins, the closed loop cycle then starts again. This project has given astounding results, so much so that it has been adopted by various institutions including The British Library, University of Winchester and Heathrow Airport in London, to name a few. In this way the Gumdrop Lt is contributing to the cleaning of the environment and at the same time offering a recycling solution, also collaborating with other manufacturers and companies from all over the world to produce useful products from the collected waste chewing gum. At Gumdrop Ltd. they recycle all types of chewing gum waste from the Gumdrop Bin containers and Gumdrops On-the-go, as well as partnering with manufacturers to provide a zero-waste alternative to landfill.

The chewing gum is then recycled to create a range of compounds for use in the plastics and rubber industries. The new compounds, GUM-TEC are then used for injection moulding, to transform the material into various objects: key rings, Reusable mugs, cutlery, coffee cups, mobile phone covers, toys for dogs, but also boots and sneakers.

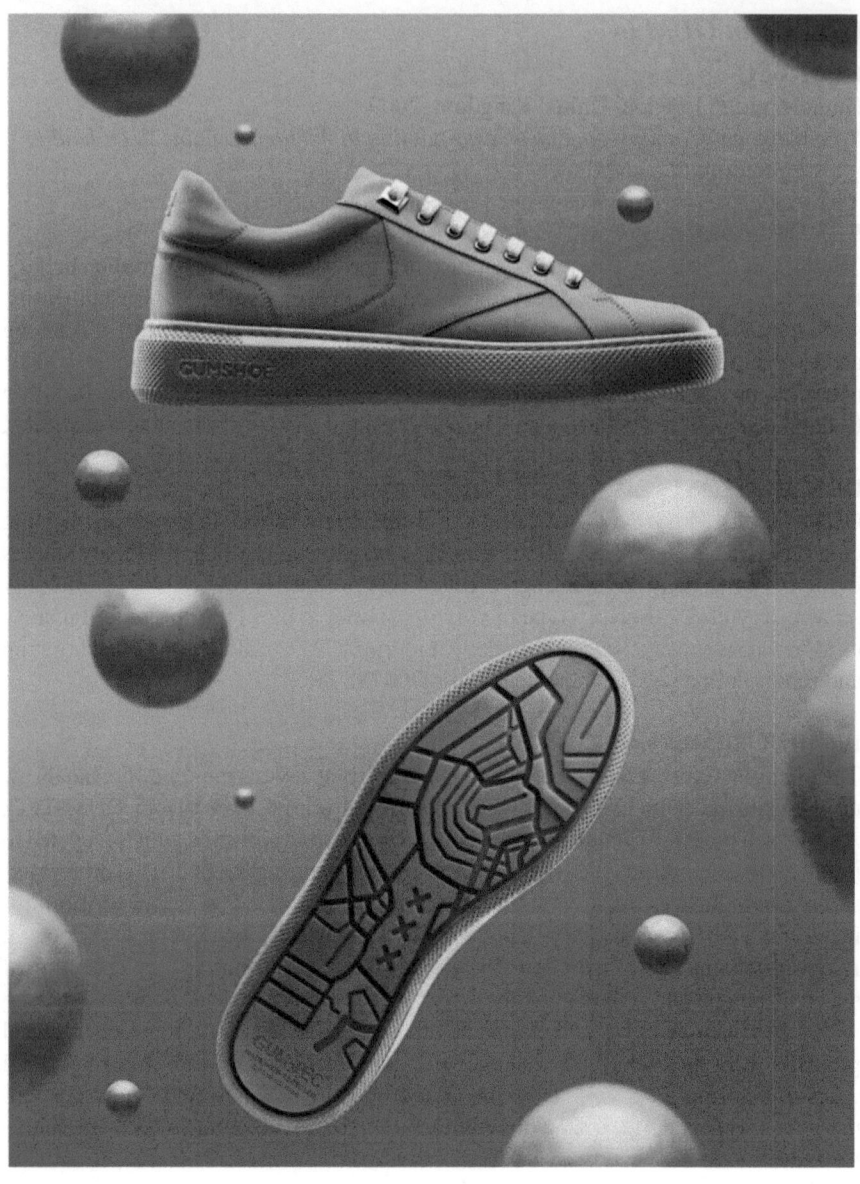

G3. | From PAPER and CARTON to

Consumption of Heritage
Sun Lee, South Corea, 2019
#Hanjipaper #crafttofactory #circulareconomy #newmaterial #eco-fabric
www.studioleesun.com

FRAMEWORK:
Korea has more than 5000 years of history and plenty of craft culture. From the 1960s to 1970s, the South Korean government pushed forward the textile and fashion industry to develop the South Korean economy and, as time passed, industrialisation and modernisation drove traditional crafts to decrease became an art, instead of being a part of everyday life. Consumption of Heritage is an invitation to reflect on the current state of fashion and to imagine solutions for sustainable materials by taking inspiration from Korean crafts, and proposing a new paradigm in production and consumption.

DESCRIPTION:
The clothes in this collection are made mainly from traditional Korean Hansan Mosi fabric and Hanji paper. They are designed for specific situations and purposes based on the relationship between the wearer and clothing, consumption and disposability and the characteristics and qualities of Mosi and Hanji. To bring this collection to life Sun Lee decided to go to SK and meet the craftsmen who are still preserving the heritage. Derived from plants, both Mosi and Hanji represent the philosophy of ephemerality, life and death, consciousness and harmony. This natural balance is realised in the clothing through a layering process, in which the combinations of and interactions between the two materials create a higher level of appreciation for quality, adaptability and context.

PRODUCTION:
Hanji paper, a traditional Korean paper made from the inner bark of the mulberry tree, has been produced for centuries and is known for its durability and versatility. Mulberry trees are cultivated and the inner bark is carefully stripped by hand from the branches to ensure the bark is intact and undamaged. The stripped bark undergoes a boiling and cleaning process to remove impurities, and then it is beaten and mulched into a pulp (manually or using machines). The mulberry pulp is then used to form sheets of paper. This is often done by hand, with artisans using traditional paper-making tools. The formed sheets of Hanji paper are carefully dried outdoors or in controlled drying facilities. The dried Hanji paper may undergo additional treatments for specific characteristics. This can include processes to enhance texture, colour, or durability. The Mosi fabric is used to create "The Transforming Coat". Embracing its modularity, it also behaves as a canvas for the Hanji garments to be connected and layered over one another. The Transforming Coat can also be combined with a long vest called Hanji feather—made with ripped Hanji paper to mimic a bird's plumage, this technique celebrates the paper's inner fibres and texture.

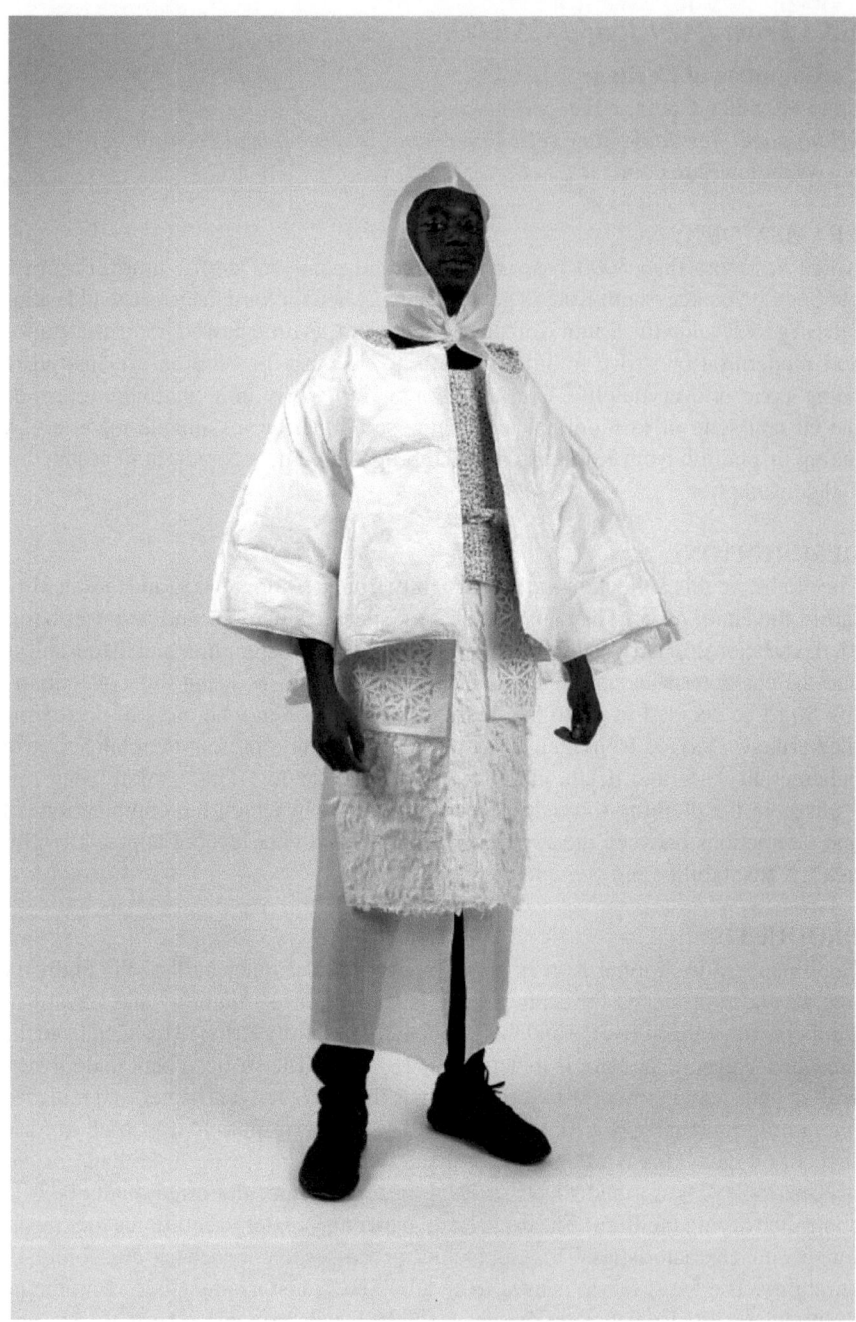

Credits: ph. Shen Yichen

G3. | *From PAPER and CARTON to*

Abitinuovi

Accademia Fiabiti, prof. Angela Nocentini, Firenze, Italy, 2018
#eggs #circulareconomy #sustainability #packaging #eco-fabric
http://www.fashionpaper.it/2010it/accademiafiabiti.htm

FRAMEWORK:

On the occasion of World Recycling Day, instituted in 2018 by the Global Recycling Foundation, the Accademia Italiana, a school of advanced training in fashion, design and photography, opened the doors of its Florence and Rome branches with an Open Day conceived in a "Reduce—Reuse—Recycle" key. The common thread, a single raw material to be reinterpreted in a creative way: paper. Old magazines were thus transformed into clothing prototypes.

DESCRIPTION:

The clothes made by the students of the Academy, coordinated by professor Angela Nocentini—Caterina Campilongo, Fulvio Caviglia, Lanzilao Federica, Stefania Maglietta, Simona Materazzini, Giuditta Scifoni, Marta Monaco, Davide Iacobucci, Serena Andrei, Chiara Roverelli, Daniela Crobe—all have in common the recycling of paper elements. From corrugated cardboard for boxes, to wrapping paper, but also old geographical maps, comics and comic books, tissue paper, egg packaging, toilet paper, labels and price tags, paper for plates under cakes, etc.

PRODUCTION:

Giuditta Schifoni, for example, proposed a version with hat and bag in this dress based on cardboard egg boxes or packaging. "The egg-boxes are also popular for their particular shape, with all those small holes alternating with just as many small bumps, like perfect and elegant folded surfaces Eggs always have been a symbol of life".

While Fulvio Caviglia, created a white and grey, horizontally striped dress made from wrapping paper and old geographical maps. "Among the thousands of paper objects we use every day is the world of geographical maps. The map is not the territory as similarly the dress is not the body'.

Lanzilao Federica and Stefania Maglietta then proposed a dress made from comic origami. "Reinterpreting the East through the West with Walt Disney and paper twine, a fusion through the use of two relevant icons: Japanese origami, the origin of which is closely linked to the Shinto religion, and the sacred value of paper, which is also testified to by the fact that in Japanese the word paper and Gods are both pronounced hami".

G4. | *From HAIR to*

Human Material Loop
HUMAN MATERIAL LOOP, Geleen, The Netherlands, 2019
#humanhair #sustainability #biotextile #circulareconomy #eco-fabric
www.humanmaterialloop.com

FRAMEWORK:
In Europe alone, an estimated 150 million kilograms of human hair waste end up in landfills or incinerators. While we typically associate human hair with being on top of our heads, beauty salons generate vast amounts of waste, yet waste management efforts in cities primarily focus on waste collection only. With the world's population rapidly increasing, proper waste management and recycling solutions are essential for the future, and transitioning to local materials and production is one of the key elements to reach a circular economy.

DESCRIPTION:
Human Material Loop is a material innovation company that develops technologies, materials and products from waste human hair to deliver high-performing products for the textile industry with a minimal footprint. The company is located at Brightlands Chemelot, the EU's largest bio-engineering facility in the Netherlands. They have developed a technology to utilise waste keratin protein fibre and develop high-performance products for the textile industry. While the most common interest in hair is focused on hair growth, hair types and hair care, it's important to recognise that hair is also a significant biomaterial primarily composed of protein, notably keratin. The unique advantages of biological fibres such as human hair have not yet been fully implemented in our product cycles. Human hair, with its abundance, non-toxicity, non-irritating properties for the skin, high tensile strength, lightweight nature, thermal insulation, flexibility and oil-absorbing capability, shows great potential for integration into our production systems.

PRODUCTION:
The human hair is a natural filamentous biomaterial and chemically, approximately 80% keratin protein is present in human hair, is the same keratin protein fibre as wool. The durability of keratins is a direct consequence of their complex architecture with extremely high molecular weight. Hair has a strength-to-weight ratio comparable to steel. It can be stretched up to one and a half times its original length before breaking. Hair behaves differently depending on how fast or slows it is stretched. The faster hair is stretched, the stronger it is. Hair consists of two main parts: the cortex and the matrix, the combination of these two components is what gives hair the ability to withstand high stress and strain. When hair is stretched under a small amount of strain, it can recover its original shape. Stretch it further, the structural transformation becomes irreversible.

Credits: photo by David van Woerden

G4. | *From HAIR to*

Contemporary Hairwork
Antonin Mongin, Parigi, France, 2020
#humanhair #sustainability #biotextile #circulareconomy #eco-fabric
https://www.infringe.com/antonin-mongin/

FRAMEWORK:
Antonin Mongin is a Design PhD and a textile crafts-designer, specialising in the preservation or revival of rare or lost crafts and know-how. The Contemporary Hairwork project (l'Artisanat d'art du cheveu coupé) aims to revive a craft practice that has been dormant since the beginning of the twentieth century. This activity consists of entrusting Antonin Mongin with his own cut hair or that of his relatives so that he can design and create customised and personalised materials or objects made from these fibres. Once cut, the hair is no longer considered as waste, but as a precious raw material charged with memorial and symbolic values linked to the owner from whom it originates. These artefacts become tangible objects of memory, as an alternative to photography.

DESCRIPTION:
Sustainability is a central aspect of the designer's work precisely due to the eco-responsible dimension of the project. The aim is to valorise waste that is often considered waste, giving them a second life in the daily lives of their owners or in the exhibitions that are created with the products created. In a subtle way, we try to change perspective and consider the result of a haircut as a precious material to be preserved and passed on. This fibre represents a fantastic conceptual starting point and material for creation.

PRODUCTION:
Weaving is one of the textile techniques proposed by Antonin to work with cut hair. He weaves them into a weft within a warp made of natural threads (such as silk, cotton or wool). Each material is unique, as it is made from a single hair donation. It is also possible to mix different types of hair if they belong to friends, lovers or different members of a family. The hair colour is not changed. Each weaving is done by hand on a traditional weaving loom.

G4./D2. | From HAIR and ALGAE to

Weaving Water
Bela Rofe, The Netherlands, 2019
#humanhair #algae #sustainability #biotextile #circulareconomy #eco-fabric
https://belarofe.cargo.site/weaving-water

FRAMEWORK:

Weaving Water is an interwoven tapestry of algae and female human hair that aims to reconnect humans to the origin of our evolution, the sea. The artwork uses textile as an agent to increase awareness about the connections between living systems.

DESCRIPTION:

Through entangling materials from the underworld (algae) with materials from the top world (female human hair), Weaving Water presents a new material ecology that can help humans feel rethreaded to the natural order. In the wake of resource depletion, there is an opportunity to co-create with, and learn from, the intelligence of these abundant bio-materials like algae and human hair. Through blending heritage techniques and biofabrication, a story about craftswomanship and ecofeminism is woven. Weaving Water aims to push traditional narratives into new speculative formats.

PRODUCTION:

Designer Bela Rofe took inspiration from the structure of seaweed farms to build and fabricate Weaving Water. She built a 2.4 m by 1.5 m timber loom to form the outer frame. The structure is held together by transparent strips of alginate and agar bioplastics cast with female human hair and dried seaweeds, encased in algae yarn. The materials are concoctions of different recipes and experiments that entangle elements of "human" and "nature" into one interconnected fabric. She used traditional techniques like manual sewing, knotting, woodwork and wet felting, alongside digital fabrication techniques including laser cutting and CNC milling.

G4. | From HAIR to

Wolfwall

Alessandra Tuseo, Italy, 2022
#dogfur #sustainability #biotextile #circulareconomy #eco-fabric
https://distributeddesign.eu/awards/entries/8553/

FRAMEWORK:
Eco-sustainable thermal-phono insulation material, which can be used both in construction and in product design, derived from canine fibres.

Twice a year dogs go through moulting, the natural process through which the undercoat is shed and allows the fur to renew itself in preparation for the warmer or colder seasons.

Annually in Italy, at least 115 million kilos of undercoat are naturally produced, of which (considering a loss of 20% of the weight after a necessary washing process) 90 million kilos of fibres could become available to be processed. Such a quantity would allow to produce up to 90.000.000 m^2 of felt.

DESCRIPTION:
In the past, before the introduction of intensive livestock farming, dog wool, also known as chiengora, was used as a substitute for sheep wool. The fibre was used also because of its hypoallergenic properties: the allergen that causes a reaction is produced by a dog's sebaceous glands, so it's not on the single strands of fur. Hence once the strands have been shed and washed, there is no trace left on the wool. Since they are handmade, the samples present imprecisions and flaws, which in turn reduce their physical efficiency.

Nonetheless, the testing has demonstrated that the product possesses some high-performing thermal characteristics. Thermal conductivity higher than average, acoustically the material possesses great sound-absorbing capacities between 800 and 4000 Hz.

Wolfwall breathes new life into a wasted resource, as well as promotes an eco-sustainable approach to upcycling, integrating itself into the tight relations between technology, enterprise and consumers as a potential, green-oriented competitor.

PRODUCTION:
The processing of these fibres involves the use of the same machinery as for the common felt: the fur is washed, dried, carded and needle punched.

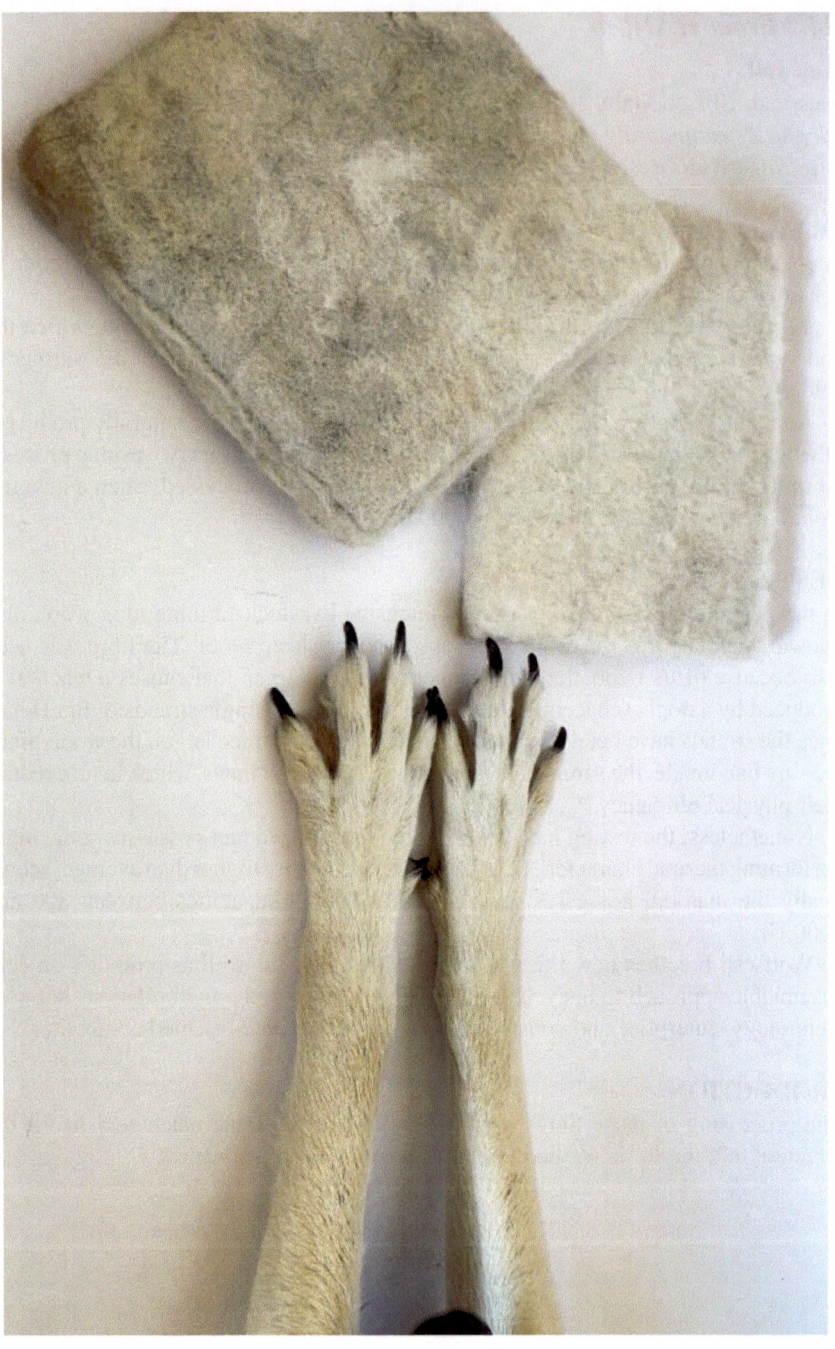

G5. | *From OTHER to*

Made In
Lena Winterink, The Netherlands, 2022
#garments #sustainability #newlabels #circulareconomy
www.Lenawinterink.com

FRAMEWORK:
The labels on our clothing do not provide any information about the origins of the materials used. Furthermore, the labels need to be cut out of the garments for a stable material-quality during the recycling process. "Made In" is a coat made entirely of the clothing labels from locally discarded garments. With this project Lena Winterink highlights the disappearance of locality from global production systems and creates a link with a new definition of locality. "Made In" started from the question of the National Museum of World Cultures to make a design in response to artefacts of the collection of the museum for the exhibition "Plastic Crush" (2023) Lena Winterink selected five objects on intuition. By no surprise, they all were related to different techniques and materials used to make wearable items.

DESCRIPTION:
She started by mapping the objects, researching the materials they were made of and the different cultures they (had) belonged to. This resulted in differences, like the materials and places of origin and time they were made, but also similarities, they are all wearable and the used materials were locally sourced. One object stood out from the others: a head from Kenia made of littered plastic. This changed the possible answers to the previous question, as found and discarded materials once made somewhere else, now also could be considered as local.

PRODUCTION:
The labels became a leftover in the recycling of textile, and here Winterink saw the opportunity for a new local material. She collected the labels from a recycling company in The Netherlands and designed a technique to attach the labels to one-another to create a new textile on which all different materials and origins were still displayed. The coat that she designed and developed, consists of over 1.300 labels from locally discarded garments and links the global mass production to a new definition of locality.

Chapter 6
Eco-textiles: New Challenges and Future Perspectives

6.1 Towards a Regenerative Agriculture: New Eco-hybrid Scenarios

Returning to the beginning of these pages, we have already pointed out that most of the current agro-industrial dynamics are responsible for more than a third of the world's anthropogenic greenhouse gases, with the consequent effects of climate change, occupy about 38% of the world's land surface and account for about 70% of all freshwater withdrawals, which is becoming critically unsustainable. Indeed, the shocking volume of resources suggests that current systems for producing food, materials and even fuel are fundamentally broken [1, 2].

To develop more circular systems for growing crops (and rearing animals), a new responsive and responsible or "regenerative agriculture" (as opposed to industrial agriculture) is needed, to make use of resources and food surpluses or waste, and to provide more resilient, diversified and multi-productive food outcomes and systems, with clear benefits for the environment, the economy and citizens [3].

We also noted earlier in this work how, in the twentieth century, food producers began to adopt a model of agriculture based on industrial processes designed to maximise the amount of food and profit that could be produced per acre.

While efficient land use was not intended to be inherently harmful, the efficiency associated with industrial agriculture was thought to be short-term, at the expense of biodiversity, human and soil health, ecological resilience and long-term food security, while also producing harmful waste products. By constantly ploughing, harvesting and replanting without integrating other organic materials such as cover crops or organic waste, industrial agriculture has lost 50–70% of the carbon it once stored.

Recent agricultural systems are therefore increasingly aware of the threat to biodiversity posed by traditional agricultural models that prioritise small numbers of plants and animals, transforming biodiverse areas into mono-pasture or single-crop fields. The increased use of fertilisers and pesticides, which leach into local waterways and kill pollinators, also affects the health of consumers.

Industrial agriculture and its waste products are damaging our environment while failing to provide long-term food security [3].

So-called "regenerative agriculture" would in fact consist of "a set of agricultural practices capable of recognising that eco-systems, environmental health and diversified mixed food and food waste treatments are central to a more sustainable and thriving system of food production" [4].

Subsequently, the concept of a new regenerative agriculture would work to actively improve ecological factors through a series of indicators such as biodiversity, soil quality, waste reduction, resource recycling and a high land capacity to sequester carbon [3].

As M. Harvey and C. Lafontaine recall, the term (coined at the Rodale Institute at the end of the 1980s) is therefore relatively new, without a very formal definition, and is in fact generally associated with the indicators mentioned above, as well as with a series of practices such as crop rotation, no-dig or low-tillage methods, and the (re)use of waste in general (such as animal faeces, dead plant matter, etc.) and food waste in particular, as a resource with multiple outcomes and multiple uses (moving towards a zero waste movement and a more circular economy by transforming waste into new resources) [3, 4].

The absence in the definition of the growing importance of the new technological and digital revolution is explained by the moment of its definition, before the impact of the new capabilities of diversified and computational production or manufacturing [5, 6].

Therefore, there is no contradiction between the two information interactive processes (with the data and with the environment), and we can speak today of a new Advanced Regenerative Agriculture (or Advanced Agriculture), where the first major benefits would be the better management of carbon sequestration and the rational or balanced management of natural and material resources, thanks to digital technologies; favouring practices with high precision in the optimisation of parameters and data indicators capable of reducing tillage processes, in addition to the precise use of compost and other organic materials instead of fertilisers; but also promoting new digital manufacturing potentials linked to second-life food, aimed at profiting from (and reusing) food waste and discards [3–5].

In this sense, Advanced Regenerative Agriculture actively increases, rather than depletes, the organic and nutrient composition of soils, plants and crops, thereby enhancing carbon sequestration. In fact, its logic is based on healthier eco-systems as a prerequisite for more accessible, innovative, effective and diversified bio-production, less dependent on large capital or industrial food distributors (which require significant inputs of fertilisers, machinery, labour and large amounts of land to operate). On the contrary, these new agro-ecological processes can be initiated with more local models—close to self-sufficient cycles—reducing the impact of mechanical effects [5, 6].

In this sense, one of the most important aspects of Advanced Regenerative Agriculture Design is that its benefits are not limited to the food industry itself, but also to other industrial sectors that deal with preferably bio-natural crops [3–8].

A key example would be the impact that regenerative agriculture is already having on biocosmetics and health products, but above all its growing momentum in biofabrics and eco-textiles and in the fashion industry (in relation to recycling in general and food in particular), which, as we have pointed out, has a serious environmental impact. For many genuinely environmentally conscious fashion companies, regenerative agriculture may be the best way to continue to produce clothing sustainably using food waste and surplus food (or food packaging residues) through innovative eco-techno-crEATive processes [1–7].

In this sense, the use of "food waste" appears as a new resource, with the use of crop surpluses, but also leaves and branches as ground cover, to encourage the growth of fungi and bacterial microorganisms, which are crucial in new textile research [3–9].

The key element is that of a new, more holistic practice in processes capable of working with agricultural crops (crops) and design ops (operations); but also with products derived from food (declinations), especially applicable in biomaterials such as bioplastics or biofabrics.

Acting through a greater positive interaction with/between the various existing environmental conditions.

This means that the most relevant feature—among the possible synergies generated between Advanced Regenerative Agriculture and the design field—is the synergetic conjugation of a new innovative equation $R + D + I + D$ (i.e. Research + Development + Innovation + Digitalisation), which is absolutely fundamental in the current and emerging economic dynamics.

Many recent studies are analysing the current flows of agricultural waste and their potential impact on the textile industry on a large scale.

The market for circular clothing is currently projected to reach \$77 billion over the next five years. Similarly, new circular innovations continue to emerge in the field of bio-fabrics and bio-textiles. For example, recent analyses show that only in South and Southeast Asia are there sufficient agricultural waste streams to support large-scale bio-production of recycled natural fibres [10].

It has already been pointed out that the combination of current industrial production processes in the textile and agricultural sectors is having increasingly disastrous effects on the environment.

More than 60% of the fibres used in clothing are derived from petroleum, putting a strain on natural resources through uncontrolled and unsustainable production.

Similarly, conventional natural fibres such as cotton (the second most widely used textile fibre) depend on intensive use of agrochemicals and water [10].

Researchers are analysing several crops to find the most suitable performative dynamics for the production of fashion and alternative biofibres, biotextiles or biofabrics; they are making practical recommendations to create, as we have said, "new value chains" based on the agricultural residues with the greatest potential, such as milk, rice, wheat and cereals, but also banana stalks, the "waste" of sugar cane and pineapple leaves, etc. [10].

At present, much of this waste is disposed of by mass burning, which causes severe air pollution. The new research proposes a kind of "new collaborative and innovative roadmap" to bring the fashion and food industries together through new, more sustainable alternative biomaterials. In order to reduce its growing dependence on fossil fuels, the fashion industry must prioritise and accelerate "its transition to circular and regenerative systems, and one way to do this is to 'create value' from food waste; to work in the field of biomanufacturing, to accelerate alternatives that tip the balance in favour of the environment, with greater climate benefits; to reduce waste and pollution, to recycle food waste and textile fibres, and in this way to generate important economic and social benefits" [10, 11].

6.2   Present and Future Eco-textile Resources and Challenges

> It takes a lot to make a garment, not just the parts we hear about - the designers, brands, shops, runway shows, parties and Instagram influencers - but also the farmers, ginners, spinners, weavers, sewers, artisans and other factory workers who produce the raw materials and turn them into our clothes.
>
> It takes water, soil, seeds, land, forests, animals, electricity, oil, chemicals, metals and other precious natural resources to clothe us. At the current rate, scale and level of technological innovation, the fashion industry is heading towards an unsustainable and uncertain future [12].

Throughout this publication, we have seen the emergence of new and promising experiences that are helping to transform not only the fashion industry, but also "the way clothes are bought".

An increasing number of new technologies and emerging processes make it possible to design, produce, use and recover products and materials in radically innovative ways: closed-loop systems in which the recovery, reuse or future recycling of consumed elements and materials is anticipated and foreseen, including the creation of new circular, ecological, economic and social processes from the outset [9–13].

By the mid-2000s, fashion had already become a huge globalised industry, with production constantly shifting to countries with the lowest wages, least regulation and least protection for workers and eco-systems: a system designed to maximise profits by producing increasing volumes and margins as quickly as possible and at the lowest cost.

In recent decades, the global fashion industry has become one of the most influential sectors in terms of industrial, cultural and financial power, worth more than $2 trillion. More than 150 billion garments are produced each year. As a result, apparel production has become the third largest manufacturing sector in the world, after automobiles and electronics. Despite years of industrial automation and technological innovation, garment production remains an intensive and labour-intensive process that employs millions of people around the world, often in poor economic and environmental conditions [12].

Today, people spend less than a fifth and buy more than twice as much as they did 20 years ago. The amount of clothing a typical average European family buys in a year generates emissions and uses unsustainable amounts of water, making clothing the fourth largest environmental impact (after construction, transport and food production), with more than 300,000 tonnes of clothing sent to landfill [12]. Millions of trees are cut down every year to make textiles, putting protected ancient forests at risk. Synthetic clothing is the main source of microplastics polluting our oceans.

Fashion and progress have always played an important narrative and cultural role. In this framework, the idea of unlimited growth through technological and economic progress has favoured a certain type of economic development that has been prolonged for decades. But now, the focus needs to be on how fashion can instead be a powerful driver in building a more sustainable and resilient, decentralised and circular future, linked to smart approaches [6–9].

Turning agricultural waste into raw material for the textile industry is a step in the right direction. But in order to move forward, we also need to take advantage of the lessons learned through constant trials, failures and "trial and error" [10].

As we have seen, circular textile innovations are increasingly emerging, using fibres based on agricultural waste to produce fabrics and accessories from different companies.

This pioneering production model can help revolutionise the fashion industry, but of course the complicity of consumers is important: it is imperative that the industry evolves alongside a consumer society that is increasingly sensitive to innovation and the environment [9, 10].

This virtuous synergy is a fundamental step towards the sustainable production of renewed and renewable fabrics. Textile production is an incredibly complex process, involving many actors, whose main objectives are fair quality and an appropriate economic and environmental impact for the production of fabrics and tissues.

Obviously, in a society that is increasingly sensitive to a fair economy and a caring environment, the ideal situation for today's textile industry is to reduce the amount of waste before and after production, thereby reducing the environmental impact, while keeping costs to a minimum. Polyester (basically a plastic) and cotton (a natural fibre) are the most dominant materials in the industry today, but both have disastrous effects on our habitat, as we have mentioned [10].

Throughout this paper, we have insisted on showing how today's textile industry is increasingly turning to nature, using natural and/or recycled fibres from plants, the sea and food waste to produce sustainable fabrics.

A large proportion of "sustainable fashion" items use blends of recycled synthetic materials (i.e. polyester) and natural fabrics, which in the near future will increasingly come from a new hyper-material concept of food or products' own food packaging (also recycled), to promote not only a more ecological circular economy but also a more creative, responsible and friendly textile production [9, 10].

In the central catalogue that makes up this publication, we have reviewed some of the basic elements involved in this process (see Chap. 5).

According to Oliva Burton, we can synthesise the most important raw materials that today begin to operatively change the textile and fashion industry [10].

Bamboo, an element that grows quickly and is cheap, making it a perfect complement to the textile industry, with a low environmental impact, despite certain toxins that tend to be used during processing.

Wool, a biodegradable fabric that can decompose in compost, with a production process that has a low environmental impact (depending on whether or not chemical dyes or toxins are added).

Hemp, which is currently used in certain types of clothing, although its use is not yet widespread. Derived from the marijuana plant, it is a versatile plant, similar to bamboo, that grows quickly and has a low environmental impact.

Chitin, a fibre derived from marine waste, mainly crustacean shells. It is cheap, versatile and easy to incorporate into a wide range of manufacturing processes, and its adhesive properties reduce the need for dyes or artificial binders.

Algae, a very versatile material with low environmental impact and interesting biosynthetic and CO_2 sequestration properties.

Banana fibre, similar to bamboo: versatile when softened and inexpensive. It is also one of the strongest fibres and can be used commercially as it is environmentally friendly and biodegradable.

Pineapple, a very interesting material that can be used to make pineapple fibres, threads and leather products through an environmentally friendly process using pineapple plant waste and natural dyes.

Orange fibre, a very important material obtained from citrus fruit waste, with great current applications in the spinning industry.

Nopal, a raw material that can be used to make soft neoskin and neoskin pop-ups.

Milk and casein, another important food source that is increasingly being used for alternatives to bioplastics and innovative production of yarns and textiles.

Coconut fibre, called coir, is made from discarded coconut shells. It is not as versatile as some of the previous ones (as it is a very rough material), but it can easily be used for bags, shoes, brushes, etc.

Corn fibre is a very adaptable and economical material that has no impact on the environment and can be used to make clothing using the dextrose present in corn fibre.

This is just a synthesis of the many current ecological (or agro-ecological) sources that we have reviewed in the previously published compendium, which are used to produce eco-textiles. Obviously, there are other interesting elements, processes and strategies that major brands are already incorporating [9, 10].

Furthermore, in the field of pigments and bio-dyes derived from food itself, many research groups are already working to replace toxic synthetic dyes with natural alternatives, ranging from specific plants and natural species to microbes and, logically, food waste [13].

Introduced in the 1860s, synthetic dyes and pigments have become commonplace in the textile industry, offering every colour imaginable. However, industrial dyeing processes also produce enormous residues and pollution.

Food by-products such as onion skins and willow bark, beet pulp or turmeric powder can be used to dye clothes, representing an untapped potential for natural dyes as new alternatives to reduce waste and pollution [13]. But natural sources of colour are not limited to vegetables or plants: fungi, mycelia and microbes offer enormous potential for the future of dyes and pigments. Bacteria can be a source of non-toxic, biodegradable pigments and a good way to help dyes adhere to biotextile fibres.

The use of bacteria in the natural dyeing process requires slower timescales to grow and feed the bacteria, although these can be easily modified to produce a wide variety of different pigments, which could help to promote greater textile production with more ranges and greater variations.

Lab-grown and/or food-derived dyes are a particularly promising future, as there is less and less land available to grow the plants traditionally used for natural dyes. Climate change is affecting our environment, making food and water more unpredictable and insecure [13].

This means that it will be necessary to divert resources to growing food. The possibility of using food waste itself to obtain pigments and also to produce indoor microbe farms expands the possibilities of non-toxic and biodegradable dyes, saving land and resources in new performative and eco-technological processes [9–11, 13].

In this sense, technology will undoubtedly be an important tool for change, and food surpluses or waste will become part of this revolution, not as simple residual elements, but as authentic raw materials of great value [6–9].

As we have described, orange threads can be made from citrus by-products and, through novel processing, form soft fabrics perfect for clothing. Bioengineered silk is being used to make fabrics stronger than steel, while the latest biotechnology allows leather to be produced in laboratories without harming animals [6–13].

Robotics and artificial intelligence are beginning to disrupt manufacturing and retailing. Some companies, such as Adidas, operate "on-demand" factories that use 3D printing, robotic arms and computerised knitting to produce trainers in as little as five hours. Automation has the potential to completely reshape the fashion workforce in the future. Textile and garment production may no longer provide employment for millions of low-skilled workers in developing countries and the big challenge will be to ensure that the benefits of these changes are shared equitably [12].

New technologies also allow the emergence of new, more decentralised business models related to the rental of clothes, favoured by applications for smartphones (which facilitate the exchange and buying/selling of second-hand clothes). In fact, the resale of clothes is expected to overtake fast fashion in the next ten years.

In addition, other types of business related to km zero products, resulting from the involvement of farmers, producers, designers and researchers or technicians (specialised in computerised and digitised processes), allow for new start-ups generated by cooperatives and associations with local implementation but with great potential for global dissemination and online sales, favouring at the same time economies

Fig. 6.1 Seaweed coloured with biomaterials. Exploring the different colour performances. Department of Seaweed. Aalto University. https://phys.org/news/2022-09-lab-grown-pigments-food-by-products-future.html

of proximity and projectivity. If these technological innovations are to be "multi-scalar" in the fashion industry, then our new "old clothes" (as well as our productive waste) have the potential to change the current dynamic for the better [12] (Fig. 6.1).

6.3   Transitioning Design: Bold Ecologies Versus Old Ecologies

We've already noted how "fast fashion", a term coined in the late 1990s as a reference to the rise of the fast food industry, would aptly describe the accelerated pace of clothing production. Fast fashion is cheap, shoddily made and designed to make style obsolete.

The pace of fashion has increased rapidly, with consumption rising by 400% in the last twenty years [14, 15].

Fast fashion's low prices are achieved through the use of low-quality materials, shortcuts in the manufacturing process and unfair labour practices. The industry thrives on impulse buying, designed to encourage mass consumption, and microtrends—or disposable pseudo-trends—designed to attract increasingly compulsive consumers.

Fast fashion, made in distant places, disconnects us from the origin of the product and masks environmental destruction, allowing rapid consumption to continue "without apparent consequences" [14, 15].

As defended by Irwing, Coward and Maione, the concept of "Transitioning Design" towards a new operational sustainability implies a broad and diverse set of design practices that have been rapidly evolving for several decades in an attempt to reduce or reverse environmental impacts and social injustices.

This transition to new, more sustainable operational logics includes design models whose main matrix is based on a new type of materiality and circularity, which includes reversibility, reuse, recycling and easy repair [14–16].

The main objective of circular design is to eliminate waste and pollution through the recirculation of products and materials and the regeneration of eco-systems.

In Cradle to Cradle (2002), McDonough and Braungart already addressed the contrast between linear models of manufacture, extraction, production and disposal and circular models inspired by nature's abundant flow of materials and energy through feedback loops [17]. Models in which technological-natural processes would be included and in which both biodegradation and carbon emission savings and resource (re)use would be fundamental.

The framework and methodology of transitioning designs are based on trans-disciplinary methods aimed at addressing complex problems from new logics far removed from the most conventional traditional habits and those of the most undesirable current ones [8–14]. Of course, intervening in complex systems that are already damaged is difficult and even unpredictable, but it is fundamental to try to promote a better future that is still in transition.

A future that, as we have said, allows the combination of new design approaches, technologies, materiality and hybrid natures towards a new type of green ecology 'contaminated' by the integration of different disciplines. The term "dark ecologies"—coined by Timothy Morton in 2007—has already translated a new interest in exploring a "new contract with nature" in relation to current "environmental factors" [18, 19].

While the adjective "dark" would encompass the "dark" condition of our current media, habitats or realities (urban and meta-urban), its sustainable complement ("ecologies") would raise the need to promote research on the "environmental conditions" themselves, capable of leaving behind old purist, bucolic, Platonic and/or Apollonian ideals: "environmental conditions" in which parameters of "noise", pollution, wear, corruption and/or hybridisation should be considered as an essential part of the processes to be addressed.

A definition based on different approaches, stimuli and research scenarios, as well as on experimental proposals, projects and operational prototypes, involved in different contexts (economic, social, material), and ultimately expressing a willingness to change and adapt, i.e. predisposed to material and formal transformations and evolutions [19].

The purity of what is still embryonic (or uncontaminated) then gives way to the warmth (somewhat cloudy) of those reactive and operational responses that are definitely tinged with a more hybrid materiality.

The longing for a genuine materiality would give way to the capacity (somewhat obscure, but performative) of the composite, the combined, the crossed, the reactive: a hybrid materiality, definitely tinged with a positive and proactive impurity.

To paraphrase Jordi Vivaldi [20]:

Following in the footsteps of thinkers such as Timothy Morton, Slavoj Žižek or Bruno Latour, the idea of these new "eco-techno-logies" not only challenges the boundaries between the organic and the synthetic, but above all offers a revised narrative for a renewed notion of matter and materials, whose legitimation is based on three basic elements: first, matter is now understood as an active platform (it not only provides the conditions for possible crossings and entanglements, but also allows the presence of natures within other natures). Second, matter is now informed: it contains a physical and/or digital structure that operates at all levels and scales.

Thirdly, matter now acts: It has an operational presence that is not merely metaphorical or symbolic, but authentically 'performative', in the field of new innovative processes.

The majority of the reflections and experiences that we have referred to in these pages can only be conceived from this new, less "essential" and "essentialist" acceptance of our old material definitions, through new paradigms in which the concepts of food, plastic, textile, leather, fabric, paper, waste or residues are fused and combined in an increasingly "hybrid" way, not for eccentric or capricious reasons, but precisely for a new, more eco-efficient and techno-operational productive and attentive will.

Dark ecologies, dirty ecologies and even grey ecologies are terms that can conjure up new creative "natufices" (nature+artifice) [21] similar to what is expressed by the proactive term "Bold Ecology"—as opposed to the redemptive and conservative Old Ecology—as it would be defined in the Metapolis Dictionary of Advanced Architecture [22] (Fig. 6.2).

Instead of the old nostalgic or pseudo-bucolic ecology (which freezes landscapes, territories and environments), we propose a bold ecology (...); no longer based on a timid, merely defensive - resistant - non-intervention or conservation, but rather on a non-imposed, projective and qualifying - re-stimulating - intervention in synergy with the environment and also with technology.

An ecology where sustainability is interaction.

Where natural is also artificial.

Where landscape is topography (and topology).

Where energy is information and technology is vehiculation.

Where development is recycling and evolution is genetic.

In which environment is field.

In which to conserve always means to intervene.

Fig. 6.2 Food fractals, geometry in nature by Daniel Otto Jack Petersen

References

1. Sommariva E (2015) Cr(eat)ing City. Agricoltura urbana. Strategie per la città resiliente. Trento-Barcelona, List Lab
2. FAO (2019) The state of food and agriculture 2019. Moving forward on food loss and waste reduction. Rome. https://doi.org/10.4060/CA6030EN
3. RTS Ed. (2022) The future of food and fabric—how regenerative agriculture is key to sustainability. In: RTS, May 2022. https://www.rts.com/blog/the-future-of-food-and-fabric-how-regenerative-agriculture-is-key-to-sustainability/
4. Harvey M, LaFontaine C (2021) What is regenerative agriculture? In: Sustainable America. https://sustainableamerica.org/blog/what-is-regenerative-agriculture/
5. Gausa M, Canessa N, with Tucci G (ed) (2018) Agro-cultures, agro-cities, eco-productive landscapes. Actar Publishers, Barcelona-New York
6. Carrabba P, Di Giovanni B, Iannetta M, Padovani LM (2013) Città e ambiente agrico lo: iniziative di sostenibilità verso una Smart City. EAI Energia, Ambiente e Innovazione 6:21–26
7. Guallart V (2014) The self-sufficient city. Actar Publishers, Barcelona-New York

8. Tucci G (2020) MedCoast AgroCities. New operational strategies for the development of the Mediterranean agro-urban areas. Trento-Barcelona, ListLab
9. Gausa M, Pericu S, Canessa N, Tucci G (2020) Creative food cycles: a cultural approach to the food life-cycles in cities. Sustainability 6487(12):1–17, www.mdpi.com/journal/sustainab ility, d. MDPI, Basel, Switzerland
10. Burton O (2003) Future fabrics—the incredible textiles you'll soon be wearing. In: The Grren Hub. https://thegreenhubonline.com/future-fabrics-the-incredible-textiles-youll-soon-be-wearing/
11. Carta M, Lino B, Ronsivalle D (2017) Re-cyclical urbanism. Trento-Barcelona, List Lab
12. Ditty S (2019) How our clothes might change the future. In: Anyone. Anywhere, British Council. https://www.britishcouncil.org/anyone-anywhere/explore/digital-creativity/clothes-change-future
13. Aalto University Reports (2022) Laboratory for grown pigments and food by-products. The future of natural textile dyes. In: Phys Org. https://phys.org/news/2022-09-lab-grown-pig ments-food-by-products-future.html
14. Cowart A, Maione D (2022) Transitioning toward the slow and long-developing experiential futures approach toward system change in fashion (Carnegie Mellon University) in Dialnet. https://dialnet.unirioja.es/servlet/articulo?codigo=8669686
15. Morgan A (Director) (2015) The true cost [Film]. Life is my movie entertainment bullfrog films. https://truecostmovie.com/learn-more/environmental-impact
16. Irwin T (2015) Transition design: a proposal for a new area of design practice, study, and research. Des Cult 7(2):229–246. https://doi.org/10.1080/17547075.2015.1051829
17. McDonough W, Braungart M (2002) Cradle to cradle: Remaking the way we make things, 1st edn. North Point Press
18. Morton T (2007) Dark ecology: for a logic of future coexistence. Columbia University Press, New York
19. Gausa M, Markopoulou A, Vivaldi J (eds) (2019) IAAC BITS advanced architecture magazine, no 9 (Black Ecologies)
20. Vivaldi J (2019) Matterlessness. On architecture, materiality and form under the allonomous condition. In: IAAC BITS advanced architecture magazine, no 9
21. Arroyo E (2003) "Natufice". In: Gausa, Guallart, Muller, Soriano, Porras, Morales (eds) The metapolis dictionary of advanced architecture. Actar Publishers, Barcelona
22. Gausa M (2003) Ecology, active (or bold). In: Gausa, Guallart, Muller, Soriano, Porras, Morales (eds) The metapolis dictionary of advanced architecture. Actar Publishers, Barcelona